The RobotBASIC
Robot Operating System
(RROS)
User's Guide

Simply the EASIEST way to build a robot!

John Blankenship

Copyright © 2012 by
John Blankenship

TABLE OF CONTENTS

Chapters

Preface

In order to intelligently handle a wide variety of generic tasks, a robot's hardware configuration must include a diverse selection of sensors and I/O capabilities. In the past, building such a robot has been a daunting task reserved for only those with significant knowledge and skill of both electronics and low-level microcontroller programming.

Now, for the price of an interface board alone, hobbyists can have a complete hardware/software solution in the form of a 24-pin IC preprogrammed with the RobotBASIC Robot Operating System (RROS).

The RROS will not only provide the physical interface needed for many motors and sensors, it will also provide the software required to seamlessly interface with all supported devices using the high-level RobotBASIC simulator-based commands and functions. This means that the RROS not only makes it easier than ever before to build a robot, it makes it easier to program one too.

Imagine being able to simply connect a compass to the RROS chip and then immediately determine your robot's orientation with a single programming command such as `angle = rCompass()`. Imagine connecting infrared or ultrasonic sensors directly to the RROS chip and being able to determine where obstacles are that might block your robot's path using `rBumper()` and `rFeel()`.

Controlling your robot is just as easy as acquiring its sensory information. When commands like `rForward` and `rTurn` are used, your robot won't just respond, it will respond intelligently, automatically ramping up and down when speeds are changed and using compass readings and wheel encoder counts when possible and appropriate.

And these examples don't begin to describe the power of the RROS. Read on to find the real power…. of simplicity.

Chapter 1

Why a RROS?

Since the very beginning, we have had a vision for RobotBASIC that promoted programming as well as building a robot. While *building* a robot is certainly a big part of hobby robotics, it is the *programming* of a robot that creates personality, intelligence, and the ability to create usable applications. This distinction is similar to the personal computer revolution of the 1970's.

Historical Similarities

In the mid 70's, if you were interested in computing, you had to build your own computer, either on your own or from one of the several kits that were available. So much of the early hobbyist's time was spent soldering and troubleshooting that very little application programming was done. Only after Apple and Radio Shack (among others) introduced fully assembled computers did the emphasis move from building computers to programming them – allowing the development of usable applications.

While fully assembled hardware was critical for this transition, there is another factor that played a huge role. It wasn't just that *you* could buy a computer, it was the fact that many people also had the *same* computer. This meant other people used the same video mapping hardware, the same joystick, the same sound generation hardware, etcetera, making it easy to share programming creations with others. Perhaps more importantly, it created a growing market for selling practical applications which further fueled the desire to create.

Games were certainly among the early applications, but simple programs to balance your checkbook soon found enough popularity to encourage the development of truly useful products like word processors and spreadsheets. In the years that followed, computing moved from an eccentric hobby to the massive, indispensable industry that it is today.

Hobby robotics may or may not follow the same course, but if such a path is to have any chance of happening, it is essential that building a robot become much easier. It is also crucial that those developing robotic behaviors and applications be able to share their work with other enthusiasts.

RobotBASIC, along with the RobotBASIC Robot Operating System (RROS) described in this manual allows both of these things to happen. Let's look first at RobotBASIC itself. It is a powerful, full-featured, general-purpose language that has the ability to communicate with, and control, external motors and sensors in a variety of ways. More importantly though, from a robot hobbyist's perspective, RobotBASIC has an integrated Robot Simulator.

The RobotBASIC Simulator

The simulator's two-dimensional appearance and easy-to-use interface can sometimes mask its true power from potential users because they do not expect something so simple to be effective. An abundance of easy-to-use sensors though, creates a platform with *far* more capabilities than most hobby robots. The plain truth is that you cannot give a robot any significant amount of intelligence unless that robot has the ability to gather information about its environment – and that means it must not only have the right types of sensors, but also an appropriate number of each type.

The RobotBASIC simulated robot has the following sensory capabilities:

- The ability to read its own battery's voltage so it knows when a recharge is required.
- An electronic compass so that it can determine its current orientation.
- The ability to move and turn with a reasonable amount of accuracy (the simulator allows you to set a level of random error so that it actions are realistic).
- A GPS system that allows it to determine its location within a specific area.
- A simple camera that allows it to identify objects of a user-specified color.
- A beacon detector that allows the robot to find and identify up to fifteen beacons strategically placed within its environment.
- Line sensors that allow the detection of drop-offs as well as the ability to follow a path defined by a line on the floor.
- A turret mounted ranging sensor that allows the robot to identify the distance to walls and other objects that might block its movement.
- Two types of perimeter proximity sensors that provide the information needed to avoid obstacles while navigating through an unknown environment.

It is important to know that the types of sensors available on the simulator, as well as the number and placement of the sensors, were chosen very carefully. Our book *Robot Programmer's Bonanza* (McGraw-Hill), demonstrates that the chosen sensors as well as their number and placement are adequate to implement a variety of basic behaviors and goes on to show that the basic behaviors can be combined to create significant applications.

Since the simulator's sensor configuration is available to everyone through free copies of RobotBASIC, it is easy to create coursework for students and contests for robot clubs. As powerful as the simulator is though, it cannot replace the joy giving "life" to a real robot.

Real Robots

Our *Bonanza* book also outlines the requirements for creating a real-world robot that can be controlled directly from RobotBASIC using the same programs used to control the simulator. The availability of such a robot would allow dedicated hobbyists to share programming algorithms, libraries of basic robotic behaviors, and even full blown applications that run on either the simulation or a real robot.

The design specification for a RobotBASIC interface with a real robot is very unique in that it allows for the use of nearly any type of sensors on the real robot as long as the information acquired is mapped into the simulator's sensory format. When this is done, a program capable of navigating the simulated robot through a cluttered environment can also command a real robot over a wireless link to do the same thing. This greatly simplifies real-world robot programming because all of the intelligence can be programmed in a PC-based high-level-language while all of the low-level details of driving motors and reading sensors are handled in a processor embedded in the robot itself. Because of the wireless link, there are no programs to compile and no files to download – when you make a change to your program you simply rerun it and watch the robot react.

The processor embedded in the real robot must provide the all the same sensory data to RobotBASIC that the simulated robot does, and it must do so, no matter what type of sensors are used on the real robot. Let's examine this idea in more detail by looking at the simulator's perimeter sensing system. The simulated robot uses four bumper sensors spaced around the robot that can detect when an obstacle is very close. There are also five sensors spaced around the front half of the robot that detect objects that are slightly further away. Finally, a turret mounted ranging sensor measures distances to objects outside the range of the proximity sensors. Proper utilization of these ten sensors along with a compass and a GPS provides more than enough information for a robot to navigate an unknown environment.

A real robot has many options for obtaining the sensory information available on the simulator. It could, for example, use snap-action switches or IR sensors or even ultrasonic sensors to determine when objects are near the robot's perimeter. A more sophisticated robot might utilize one or more cameras to determine if nearby obstacles exist and how close they are to the robot itself. Our objective was for a sensory system could gather the data for you no matter what type of sensors were available on the real robot. After the data was gathered, we wanted the system to analyze it, organize it, and translate it into a form compatible with RobotBASIC's simulated robot. Imagine the power of such a system – not only would it make it easier to program a robot, it would promote creativity and collaboration because it would make it easy to share programming concepts and designs with others.

Creating such a system is not a trivial task. Today's robot enthusiasts have many sensory options available to them. Some sensors are digital while others are analog and both types can sometimes require specific timing or pulsing sequences for accurate readings. More sophisticated sensors often have a complicated serial I^2C interface in order to handle the communication necessary to both control the sensor and gather information from it.

Fortunately, the manufacturers and/or the retailers of sensors usually make detailed information available on how to use their products. Often there is even example code provided for specific processors. This means that anyone with a reasonable, sometimes even a modest, background in electronics and low-level programming can handle the interfacing one specific type of sensor. Unfortunately, the task of interfacing numerous sensors so that they can all work together without conflict can be daunting. If you further consider that most of the sensory interactions must occur while the robot's drive motors are being controlled, the problem becomes even more complex.

If such a system was available, it would satisfy the needs of most hobbyists, but we still don't think that is enough. We feel that a complete solution should also have the ability to communicate with external processors so that advanced hobbyists or even manufacturers of robotic parts can create their own subsystems and seamlessly integrate them to work with all the other features. It should be possible, for example, for an advanced user to create an external vision system that extracts meaningful sensory information from the images it collects and then make that information available through the standard simulator commands and functions.

The RobotBASIC Robot Operating System
Such a system may seem impossible to achieve, but after thousands of hours of work, we have implemented a solution in the form of a RobotBASIC Robot Operating System (RROS). Just as the Window's operating system manages the various devices utilized on a PC, our RROS manages your robot's resources. The programs that run on your PC do not care what video card you use or what brand or size of hard drive you have. They do not care if you have a standard mouse or a touch pad. The Window's OS will automatically format an application program's output to a video monitor, or an LCD screen or even a video projector even though each has a different physical interface. Applications will receive identical signals whether you use a touch pad, a mouse or even a touch screen. Our RROS will provide the same type of service for your robot. It will extract data from your robot's sensors, and it will control your robot's motors – no matter what type of compatible sensors and motors you use.

Build Your Robot YOUR Way
This means you can build your robot *your* way. It won't matter if you use DC motors or continuous rotation servomotors to drive your robot, because the RROS will automatically generate the signals *your* motors need. The commands used to control the simulator can be used to control a real robot powered by either DC motors or servomotors. Small DC motors (up to 1 amp) can be driven directly from the RROS chip without additional hardware. If larger motors are needed, the RROS can control standardized external hardware so that even 30 amp motors can be used.

Your robot can utilize an electronic compass and a beacon detector as well as numerous types of IR sensors and ultrasonic sensors to handle object detection, distance ranging, line sensing, even wheel encoding – all without the need for any low-level programming – in fact, you won't have to worry about the operation of the sensors at all.

The RROS is distributed as a 24 pin IC that allows many sensors and motors to be **directly connected to the chip** without any other parts. When additional parts are required, care has been taken to minimize what is needed. This means that using the RROS is VERY economical because most of the time you DO NOT have to purchase separate I/O boards – the RROS handles everything

Beginners
The end result is that nearly anyone can now build a robot with all of the sensory and drive capabilities of the RobotBASIC simulator and control that robot directly from the RobotBASIC environment over a wired or wireless link. You can build anything from a small inexpensive robot with only a few fundamental capabilities to a large life-sized robot (complete with one or two arms) with wheel encoders, speech, vision, a positioning system (GPS or LPS), and more.

Building a robot has never been so easy. Early chapters in this book will discuss the *many* options available to you, explain what motors and sensors you can use and how to connect them to the RROS chip. Programming a robot is easier now too. Algorithms can be tested and debugged using the simulator, then immediately tested on a real robot. As long as your programs control the robot's behavior using sensory data (rather than open-loop control), both the simulator and the real robot should respond in very similar manners. As the applications to be handled get more complex, programs developed on the simulator can require modification to make them ready for real-world situations, but the overall development process can still be far shorter than developing real-world behaviors from scratch. The time to develop applications can be even further reduced if you prepare a library of basic behaviors, as discussed in Chapter 16.

Robot Building Simplified

Let's look at an example just to show how easy building a robot can be. The simulated robot has five proximity sensors around the front of the robot as shown in Figure 1.1.

Figure 1.1: The simulated robot has five proximity sensors.

Assume we wanted to build a real robot that uses five of the digital infrared sensors shown in Figure 1.2. These units are inexpensive and can be purchased from companies such as Pololu.com. Pololu also sell low-cost motors and wheels making it easy to build your base platform.

Once built, you could mount the infrared sensors (Figure 1.2) around the front of the robot as shown in Figure 1.1.

Figure 1.2: This small IR sensor detects objects up to four inches away.

Once everything is physically constructed, the wiring necessary is shown in Figure 1.3. Notice that you also need a Bluetooth transceiver for the robot (as well as a Bluetooth USB dongle for your PC – unless your PC has built-in Bluetooth) and a 5V regulator. The extra items for the robot are shown in Figures 1.4 through 1.6.

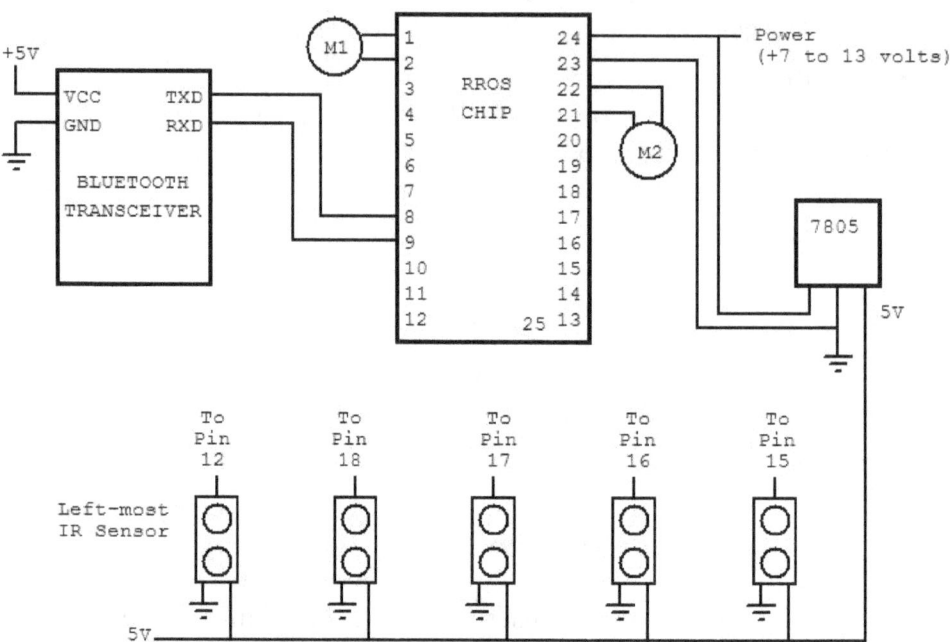

Figure 1.3: The IR sensors as well as the motors and other necessary components wire directly to the RROS chip.

Figure 1.4: This Bluetooth transceiver is available from our webpage – and others are available from many vendors.

Figure 1.5: This regulator reduces the battery voltage to 5V to power
sensors and various other components needed to build a robot.

Figure 1.6: When you plug a Bluetooth adapter into your PC's USB port, RobotBASIC can communicate wirelessly with the transceiver in Figure 1.4. There are no wires to your robot and no programs to download. Just run your RobotBASIC program on the PC and the robot responds.

Building a robot with the RROS chip really is this easy. And programming a robot is easier too. You can read the proximity sensors using the command rFeel and move the robot with rForward and rTurn. The following statement, for example, would move the robot forward only if none of the sensors are triggered.

if NOT rFeel() then rForward 10

Just an Example
Remember, this is just an example configuration. The RROS chip supports both ultrasonic and IR ranging sensors, a digital compass, beacon detection, wheel encoders, line sensors and much, much more. It can handle both small and large DC motors and servomotors. The RROS provides the physical interface for most of these devices as well as a software interface that makes it easy to read the sensors and control the motors. It is important to realize that all sensors have advantages and disadvantages for different situations. Read this entire manual to help you determine what sensors might be best for your situation.

The RROS can seem a little complicated when you first start using it because there are so many options. Different kinds of sensors have to be connected to different pins because some sensors are analog and others are digital. This manual comprises many pages because it address how to wire so many different types of sensors with various optional configurations. Once you decide what sensors (and motors) you wish to use though, the wiring is usually as simple as that shown in Figure 1.3.

And perhaps more importantly, if you have learned to program a robot using our simulator, nearly all of the simulator commands will work with your real robot.

This manual also spends many pages explaining how to calibrate motors and sensors so they are much easier to use. Again, reading about such calibrations can seem overwhelming at times but know that it is usually far easier than it seems. Plus, in most cases calibrations have to be done only once. Let's look at an example.

The command rForward 40 makes the simulated robot move a distance equal to its diameter. You will want to calibrate the RROS so that it can automatically move the REAL robot a distance equal to IT's diameter when you issue the same command. Once these calibrations are done for a specific robot, you never have to do them again unless you make physical changes.

The important thing to remember is that the RROS has been designed to allow you to utilize a wide variety of motors and sensors, and yet do so with minimal effort.

Advanced Users

Even though everything is easier, don't think for a moment this is a system only for beginners. The RROS has been designed to interface with subsystems that advanced users can design (such as the camera interface mentioned earlier). This means the RROS can grow with you when you need it to. Later chapters in this book will supply information on how to create custom interfaces for nearly any type of sensor and you will see details of how RobotBASIC can utilize the RROS to control a robot arm.

Our next step is to connect the RROS chip to your PC and ensure that it can communicate properly. Once that is accomplished we will move on to interfacing motors and sensors with the RROS chip and learning how to best utilize them from the RobotBASIC environment. Finally, we will see how the RROS can communicate with external expansions (such as an arm controller) giving you the ability to not only expand the RROS's capabilities but the ability to create totally customized capabilities to handle things that perhaps only your robot needs.

Chapter 2

Communicating with the RROS

The purpose of this chapter is to interface the RROS chip to your PC and confirm that it can communicate with RobotBASIC. The RROS chip comes without the pins soldered to it, as shown in Figure 2.1. This allows you to eliminate the pins and solder directly to the connecting points enabling ultra small robot projects. For most situations though, you will probably want to solder the pin strips to the chip, also as shown in Figure 2.1. This allows the chip to plug into a standard socket or even a solderless prototyping breadboard. The RROS chip itself is a Baby Orangutan processor, but it has been modified for our use.

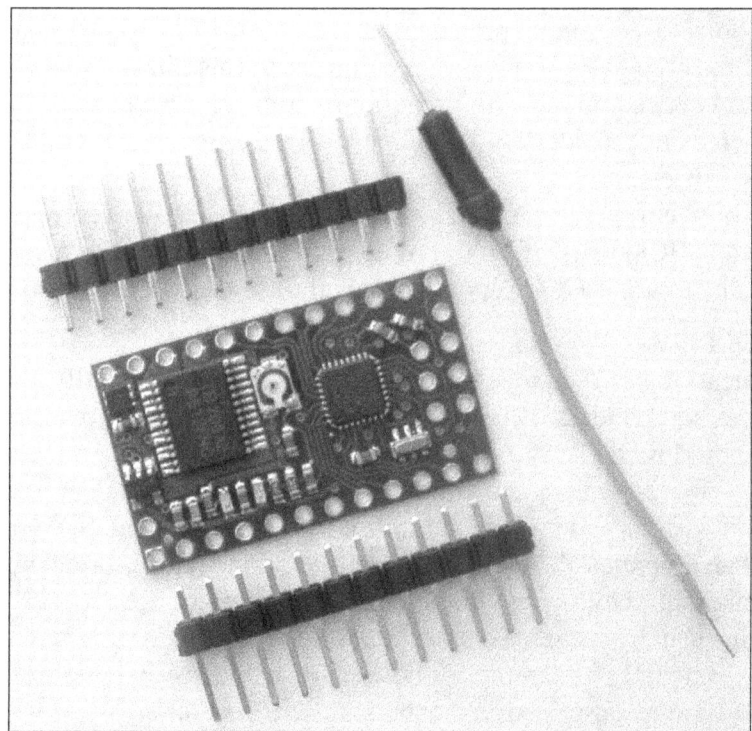

Figure 2.1: First, solder the pin headers to the RROS chip.

It is important to get the pins aligned correctly before soldering. An easy way to do this is to place the pins in the breadboard (see Figure 2.2), then place the chip onto the pins, then solder.

Figure 2.2: Use the breadboard to align the pins for soldering.

Although the RROS chip assumes the standard 24-pin footprint, the RROS actually needs 25 pins to handle all its functions. The "extra" pin can be found beside Pin 13. You can see it easily in the upper right corner of the chip in Figure 2.1. We will refer to this pin as Pin 25. Notice also in Figure 2.1 that a short wire (complete with breadboard connector pin) has been provided to make connections easy when using a solderless breadboard. This wire comes with your RROS chip and should be soldered to Pin 25

The RROS chip you receive is designed so that it cannot be read or reprogrammed by the user but it can be upgraded by RobotBASIC with any future enhancements for a small handling charge. The six pins next to Pin 25 are used for updating the chip, so do not solder anything to them.

The Communication Link
The typical communication between the RROS and RobotBASIC is usually handled over a 9600 baud wireless link, characteristically Bluetooth or Zigbee (but any serial wireless devices with similar capabilities should work).

A wireless link is very convenient and certainly fast enough for many applications. It is also worth mentioning that you can use a *wired* serial link (perhaps from a USB Serial PC dongle). A wired link is faster than a wireless link and can improve performance for advanced applications. If you use a wired link, then the PC (laptop, netbook, etc.) running RobotBASIC must reside in the robot itself. While this usually means the robot must be relatively large, there are many advantages to this approach. For example, RobotBASIC can directly handle voice recognition, voice synthesis, and vision as described in our books *Hardware Interfacing with RobotBASIC* and Arlo: *The Robot You've Always Wanted* (available Summer 2015 on Amazon.com).
IMPORTANT: Wired links MUST be 5 volt, TTL levels, NOT the standard RS232 ±12 volt levels.

We will use Bluetooth communication with the robots in this book, with the PC end being handled by an Abe USB dongle as shown in Figure 2.3. The RROS end of the communication can be handled by any compatible Bluetooth transceiver. We have had no problems with transceivers on the RROS end, but have found some incompatibilities with the PC-side USB transceivers, especially those that use their own "enhanced" drivers instead of the standard Window's drivers. The Abe dongle has always performed flawlessly no matter what transceiver we connected it to, so it is our adapter of choice. You should not have problem with other adapters in most cases, but it is important we mention the possibility. We will discuss a way to test your interface shortly.

Figure 2.3: We have found no incompatibilities with the Abe Bluetooth USB transceiver.

Testing the Interface

Before we try to communicate with the RROS, let's test the wireless interface itself to ensure that data can transfer without error. There are only four connections to a typical Bluetooth transceiver (refer to the documentation for your particular device).

Two pins are generally used to supply 5 volts (VCC) and ground (GND) to the transceiver. Two additional pins receive data (RXD) and transmit data (TXD). For testing purposes, we will just connect RXD to TXD so that anything received by the device (from RobotBASIC) to be transmitted back (to RobotBASIC). Figure 2.4 shows how to apply these connections using a solderless breadboard. Remember, the power terminals must connect to a 5 volt supply. If you do not have such a supply, three standard C-cell batteries in series should be close enough, especially for testing. Later chapters will discuss better ways of producing 5 volts for the circuits that require it.

Figure 2.4: Tie the wireless transceiver's transmit and receive pins together for testing.

The Testing Software

The program in Figure 2.5 shows a simple RobotBASIC program for testing the communication interface.

```
SetCommPort 49  // use YOUR port address here
tot=0
for i=0 to 255
 SerialOut char(i)
 repeat
  SerBytesIn 1,x,n
 until n=1
 print i;ascii(x);
 if i=ascii(x)
  print "GOOD"
 else
  print "BAD"
  tot = tot+1
 endif
next
print "Total errors = ",tot
end
```

Figure 2.5: This program tests the communication interface.

When you insert your USB Bluetooth transceiver for the first time, it should automatically install the appropriate Window's driver. You need to pair it to your remote Bluetooth transceiver using the instructions that came with it (typically through the Bluetooth icon in your system tray or on the Window's Control Panel). Generally, you will have to enter the key (password) that was specified in your device's documentation. Once the pairing is complete, the devices will connect automatically each time the two devices see each other.

Once paired, ask Window's to show you the Bluetooth Devices available (again using the Bluetooth Icon) and you should see a window similar to Figure 2.6 which provides you with the actual Port Address Windows assigned to YOUR device. You will need this address to establish communication between the two transceivers.

The test program in Figure 2.5 starts by initializing the serial port assigned to your Bluetooth connection. A for-loop is used to send all possible byte combinations over the serial connection. Since the transmit and receive pins on the remote device are tied together, the data sent out will be immediately transmitted back. When this byte is received by RobotBASIC, it is printed and compared to the original transmission. If communication is working, you will have no errors.

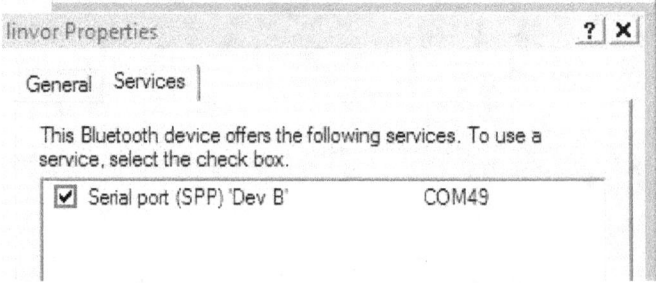

Figure 2.6: Window's can provide the Port Address used for your Bluetooth communication link.

Connecting to the RROS

As mentioned earlier, using a solderless breadboard is one of the easiest ways to wire your circuits. We have created numerous robots for testing the RROS using this simple technique (see Chapter 15). Even if you eventually wish to permanently solder all the connections for your robot, it is certainly suggested you utilize the breadboard approach until you have everything working exactly the way you want it. We were always swapping sensors and trying different configurations with our prototypes so, to make rewiring of circuits easier, we generally used wires that were longer than necessary which often caused our circuits to appear messy and disorganized. Since most people will not need to use a variety of sensors, it should be easier to create a professional look even with a breadboard.

Figure 2.7 shows a schematic diagram showing how to connect your transceiver to the RROS chip. Notice that our recommended power requirements for the RROS chip is between 6 and 12 volts. Figure 2.8 shows the actual connection between the RROS chip and the transceiver using a breadboard.

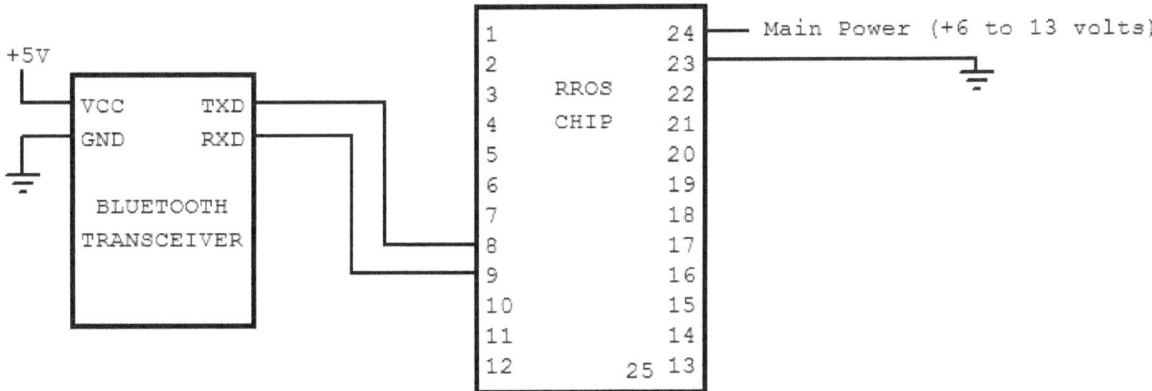

Figure 2.7: This schematic diagram shows how to connect a transceiver to the RROS chip.

Figure 2.8: When implemented, the schematic shown in Figure 2.7 will look like this. **Note**: No power is shown to the RROS chip.

Testing the RROS

Now that we have the communication aspect of our circuit implemented, we can move on to confirm the RROS is operational. To make using the RROS as easy as possible, we have provided an include file called RROScommands.bas (available from the RROS TAB at RobotBASIC.org). You should include this file in your programs as shown in Figure 2.9.

```
#include "RROScommands.bas"
main:
 gosub InitMyRobot
 // your programs will be written here
end

InitMyRobot:
 rCommport 49  // Use your Port address
 rlocate 0,0
 gosub InitCommands //found in the Include file
 // statements will be added here
 // throughout the book
return
```

Figure 2.9: This is a template for programs that you write for controlling a RROS-based robot.

The program shown in Figure 2.9 is incomplete. It is only a basic template that you should use when writing *any* program to control a RROS-based robot. The program starts by including the RROS command file that sets up many constants that will make RROS programming easier.

You should also see that there are two major sections to the template. There is a main program that normally will control your robot. There is also a subroutine, that in this case is called

InitMyRobot. As we proceed through this book we will show how to customize this subroutine to initialize all the parameters appropriate for *your* robot. Let's discuss this idea before moving on.

Your Initialization Subroutine

Remember, the RROS can handle many different types of motors and sensors. This means that before you start using it to control your robot, you *must* tell it what motors and sensors you are using. The commands to do this will generally reside in your InitMyRobot subroutine. The subroutine will also contain commands that can calibrate your particular hardware so that your robot will operate as expected. All of these commands will be explained in detail as we proceed through the book.

Naming Your Routine

The name of your initialization subroutine is entirely up to you. If you have several robots using different types of motors and/or sensors, then you will want to have several initializing subroutines, one for each of your robots. This will make it very easy to use the same program to control any of your robots (or even the robots of others at a club meeting, for example). If you create separate include files for each of your robots then any program you write can control any of your robot by changing only a single line of code that calls the appropriate initialization subroutine. This may seem complicated, but it will become clear as we proceed through the text.

Often, in this text, we will make references to statements that should be added to your main program or to your Initialization subroutine. Early on, we will show these changes in detail, but as we proceed we will assume that you know to add the commands properly. This will allow us to minimize the space needed for program listings because we will not be repeating code that has already been shown and explained.

Now that you understand how RROS programs will be organized, let's create a test program to demonstrate that the RROS is functional. The test program will also serve to show you a simple example of commanding the RROS to perform tasks for us.

Making Sounds

Sometimes it is valuable for your robot to be able to make some simple sound effects. For example, it might issue some beeps to let someone know they are in its way or it might want to play a little tune to celebrate when it has accomplished some goal.

Because of this, we gave the RROS the ability to make sounds at the *remote* robot. This ability was never intended to produce high quality music or any significant audio response such as a voice. That can be done from the PC directly, and advanced robots needing such capabilities are probably be better served with an embedded PC anyway.

That said, we felt it would be nice to have some limited sound generation capabilities built into the RROS and we can use them now to confirm that the RROS is working.

Of course we will have to connect a sound transducer to the RROS chip as shown in Figure 2.10. We will actually alter this interface in a coming chapter, but for now, this will work fine. You may use most any piezo buzzer. Be careful to match the positive/negative markings (if any) on your buzzer to those in the schematic.

Try turning on the power to the RROS chip with the buzzer attached. You should hear a short tone indicating that the RROS chip has become operational. If you do not hear this tone, turn off the power and check your wiring carefully.

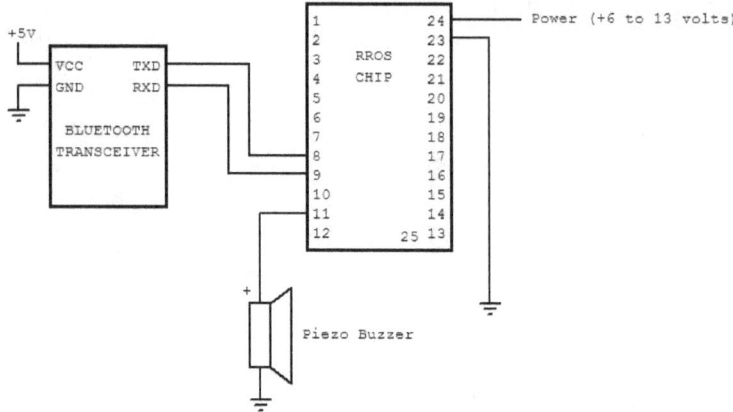

Figure 2.10: Adding a piezo buzzer lets your robot make sounds.

Now let's make the remote robot produce a simple tone. Add the following line to the Main program shown in Figure 2.9 (add the new line just before the END statement).

rCommand(PlaySound,LowTone)

When you run the modified program you will hear two sounds. The first will be the same sound generated when the RROS powers up. This sound also occurs when the RROS is initialized with an **rLocate** command. The second sound heard will be a lower tone produced by the **rCommand**. You can verify this by placing the command **delay 2000** before your **rCommand**. You will then hear the power up tone, followed by the low tone about two seconds later.

We will use the **rCommand** extensively throughout this book to issue special commands to the RROS. The **rCommand** will always have two 8-bit parameters. The first is the command code and the second is used to qualify what is to be done. In this example, the first parameter is telling the RROS to play a sound, and the second parameter specifies what sound to play. Try changing the parameter **LowTone** to one of the options shown in Figure 2.10.

Parameter	Description
Blip1	drip/blip sound
Blip2	drip/blip
InitTone	startup RROS tone
LowTone	low tone
BeepBeep	two quick beeps
BeepBeepBeep	three quick beeps
Phasor	a phasor sound
Siren1	a type of siren
Siren2	another siren
Siren3	still another siren

Figure 2.10: These are the standard sounds for the RROS.

All of the parameters in Figure 2.10 are simply numeric values. You can see the actual numbers that they represent by examining the assignments statements in the file **RROScommands.bas**. In fact, all of the numeric codes used in **rCommands** are summarized in a quick-reference format in Appendix A.

Playing Music

You can also play a particular note (C, D, E, F, G, A, or B) in one of three octaves using a specifying parameter such as **MidC**, **HighA**, or **LowG**. You can determine the length of each note by ORing (|) it or ADDing it (+) with a length designator (**Double**, **Whole**, **Half**, or **Quarter**). This means you can use RobotBASIC's **DATA** command to create the notes for a song as shown in Figure 2.11. This song is actually included in **RROScommands.bas**, so you can try it without having to type it in.

```
Data Birthday; MidC | Quarter, MidC | Quarter
Data Birthday; MidD | Half, MidC+Whole, MidF+Half
Data Birthday; MidE+Half, Pause+Half
Data Birthday; MidC+Quarter, MidC+Quarter, MidD+Half Data Birthday; MidC+Whole, MidG+Half, MidF+Half
Data Birthday; Pause+Half
Data Birthday; MidC+Quarter, MidC+Quarter, HighC+Half Data Birthday; MidA+Half, MidF+Half, MidE+Half
Data Birthday; MidD+Whole, Pause+Half
Data Birthday; MidB+Quarter,MidB+Quarter,MidA+Half
Data Birthday; MidF+Half,MidG+Half,MidF+Whole
Data Birthday; 0  // each song must end with a zero
```

Figure 2.11: These notes play Happy Birthday. Notice that
you can use either the + or | symbol to combine terms.

The subroutine needed to play a song is also included in **RROScommands.bas**. You can play the **Birthday** song by adding the following commands to your **Main** program.

```
mcopy Birthday,CurrentSong
gosub PlayMySong
```

The subroutine **PlayMySong** will play the song stored in the array **CurrentSong**. The **mcopy** line (above) copies the notes stored in the array **Birthday** into **CurrentSong**. The second line calls the subroutine **PlayCurrentSong** to actually play the notes.

Remember, the sound abilities of the RROS are not intended to be of high quality, but they can provide your robot with some basic sound effects to quickly and easily give it some personality.

Play with the sound commands to get comfortable with using rCommands with the RROS. When you are ready, move on to the next chapter where we will start controlling your robot's motors to produce movement.

Chapter 3

Small DC Drive Motors

No robot is complete without some movement capabilities, and that generally means motors. The RROS has been designed to handle the requirements of almost any application. Small robots can be made that are powered by either DC motors or servomotors. Large robots can be powered by large DC motors that require up to 30 amps each. Let's look first at the small DC motor option.

Small DC Motors
Figure 3.1 shows the robot we used for prototyping the RROS DC motors routines. It also demonstrates several sensor options, but that will be the subject of a later chapter.

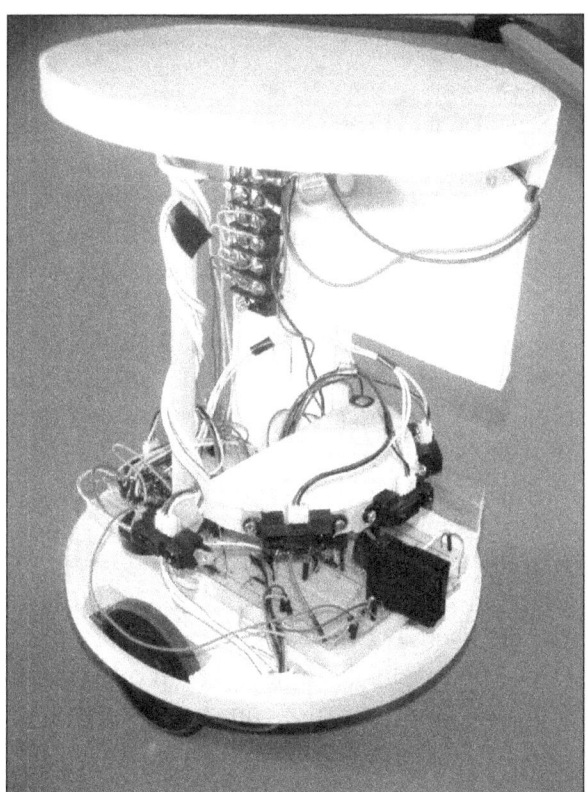

Figure 3.1: The drive system for this robot is two small DC motors.

The motors used are shown in Figure 3.2, which is a bottom view of the robot. The motors have a 200:1 ratio gearbox with a 90º output shaft. They can be purchased from Pololu.com (item #1120) which also offers wheels that mate directly to the motors. These motors are ideal because their low current requirements allow the RROS chip to drive them directly. More on this shortly.

Figure 3.2: Small DC motors provide motion for the robot in Figure 3.1.

Figure 3.2 also shows one of two 3-gang battery holders used to provide power for this robot. The six batteries in series provides power for the RROS chip itself (Pins 23 and 24 as shown in Figure 2.7 in Chapter 2). Figure 3.2 also shows a hand-made encoder disk on the wheel (the encoder itself is mounted to the right of the wheel). The encode system will be discussed in Chapter 7. The motors are driven from the main voltage applied to the RROS chip.

The robot's main body parts are made from foam board (which is both strong and light weight) available from most craft stores. The motors are attached to balsa wood blocks (which are glued to the foam board) with small screws.

It is possible that you do not have your robot assembled at this time but that is not a problem. You can test your motors by simply connecting them to the RROS chip. Connect the two wires for your robot's LEFT motor to the RROS pins 1 and 2 and the wires for right motor to pins 21 and 22. Motors connected directly to the RROS chip like this must be small (drawing less than 1 amp each). Larger motors will be discussed in a later chapter.

You may need to reverse the connections to either or both of your motors based on how it is mounted etc. Reverse the connections if the motor runs backwards when you expect it to run forward.

Any motor (large DC, small DC, or servo) will not receive power unless the RROS has been properly told of its existence. This is done with an rCommand as shown below. **Note**: This command should be placed in the InitMyRobot subroutine discussed back in Figure 2.9. **NOTE:** The MotorSetup should be done FIRST before sensors are setup.

rCommand(MotorSetup, Param)

The value of the lowest three bits in the parameter Param tells the RROS which motors are being used. When we are using small DC motors as in this example, the parameter should be zero; there are other codes for other motor types.

To make the RROS easier to use, the include file mentioned earlier sets up variables that represent various RROS options. For example, instead of using a fixed numeric value for Param, you can just use SMALLDC line this.

rCommand(MotorSetup, SMALLDC)

If your robot does not use wheel encoders (these will be discussed in Chapter 7) then this one parameter is all you need. If your robot has encoders, then you can OR or ADD another parameter (ENCODERS) as shown in the two examples below.

rCommand(MotorSetup,SMALLDC+ENCODERS)
rCommand(MotorSetup,SMALLDC|ENCODERS)

Either the + sign or the OR symbol (|) may be used. Since we have not discussed wheel encoders yet, we will proceed assuming your motors do not have encoding capabilities.

Controlling the Simulator
For those readers that might not be familiar with the RobotBASIC simulator, let's start with a simple program that controls the simulated robot. Enter the program shown in Figure 3.3 into RobotBASIC.

```
main:
 rLocate 400,300  // initializes the robot
 rForward 120
 rTurn 90
 rForward 120
end
```
Figure 3.3: This program moves the simulated robot.

If you run the program in Figure 3.3, the simulated robot will be initialized near the center of the screen and then move forward a distance equal to three times its diameter (the default simulated robot is approximately 40 pixels in diameter). The robot will then turn right 90° and move forward another 120 pixels before stopping. **Note:** Chapter 16 provides detailed RROS programming examples.

Controlling the Real Robot
Ideally, the program in Figure 3.3 can be used to control the real robot and have it respond in a very similar manner to the simulated robot. In the long run, this is certainly an achievable goal, especially when the robot's movements are being governed by information obtained from sensors (closed-loop control). For example, we could program the simulated robot to move forward until it finds a wall, then use its sensors to "feel" its way along the wall. If the same program is used to control a real robot, then the results will be very similar as long as the sensors used on both robots have similar placements etc.

We can modify the program in Figure 3.3 so that it controls a real robot by using the techniques discuss in Chapter 2, Figure 2.9. An example modification is shown in Figure 3.4.

```
#Include "RROScommands.bas"
Main:
 gosub InitMyRobot
 rForward 120
 rTurn 90
 rForward 120
end
```

Figure 3.4: This modified version of Figure 3.3 will control the real robot instead of the simulator.

For the most part, the change made in Figure 3.4 compared to Figure 3.3 is that the robot is being initialized by the InitMyRobot subroutine instead of just an rLocate statement. Of course you must have the InitMyRobot subroutine (as discussed in Chapter 2) and include the RROScommands.bas file.

Since programs like the one in Figure 3.4 do not use sensory information to affect the robots movements (open-loop control) the actions of the real robot will not necessarily mimic the simulator accurately. For example, the real robot might not move in a perfectly straight line when looking for a wall. While it is not essential that the real robot and the simulated robot track each other's movements exactly when open-loop control is used, it is valuable to obtain some level of similarity in order to increase the value of developing programs with the simulator. This can be accomplished in two ways.

First, the simulator can be made to react much more like a real robot. The command rSlip 10, for example will add up to 10% random error to the simulated robot's movements. This let's you use the simulator to create algorithms and behaviors that better deal with real-world situations.

Second, we can fine-tune the real robot so that its movements have as little error as possible. In Chapter 7, we will examine how wheel encoders can provide feedback so that we have a closed-loop system that can help keep the robot moving a straight line and make turns more accurately.

In many cases though, the additional expense and work of adding wheel encoders is not necessary. As long as the real robot's open-loop movement is reasonably close to that of the simulator, then sensory-based behaviors should operate properly. Because of this, we added commands to the RROS to allow the user to fine-tune the robot's open loop movement. Let's look first at how we can ensure that the robot moves in a relatively straight line when asked to do so.

Improving Open-Loop Control

The reason a robot might not move in a straight line is that the two drive motors are not evenly matched. If one motor is more efficient electrically or if that motor has less friction, then that motor will turn slightly faster than the other motor (even when they are told to move at the same speed) causing the robot to drift to one side when it is commanded to move forward (or backward). If, for example, your robot drifts to the left when it moves forward, you can use the following command to slow down the right wheel by 5%.

rCommand(SetReducForwRight,5)

Similarly, if the robot was drifting hard to the right, we could slow the left wheel by 10% with this command.

rCommand(SetReducForwLeft,10)

Two other rCommand parameters (SetReducBackRight and SetReducBackLeft) can be used to slow down a designated wheel when the robot is going backward. The ability to establish different percentages of slowdown for forward and backward movements is important because many motors have slightly different physical characteristics depending on their direction of rotation. Notice that these commands only let you slow down one of the motors – never speed one up.

You should experiment with your robot and determine exactly what fine-tuning is needed to make your robot move in a relatively straight line. The commands you determine to be necessary should be placed in the InitMyRobot subroutine discussed in Figure 2.9, Chapter 2. This will force your robot's movements to be fine-tuned every time a program is run. If you have several robot's you should have separate initializing routines for each of them (each aptly named for the corresponding robot). The ability to #include the appropriate initialization routine in your programs makes it easy to use any program you write with any of your robots.

Even if your robot uses wheel encoders, the RROS's ability to compensate for drift will be enhanced if the normal operation of the motors has been balanced as discussed above.

Fine-Tuning Turns

You can also control your robot's open-loop turning movements. If you use the command rTurn 90, for example, the simulator turns 90° to the right. If your real robot does not turn the proper amount you can vary two parameters to make this happen as shown below.

rCommand(SetRotationTime, 10)
rCommand(SetSlowDownSpeed, 30)

If your robot is turning too much (more than 90°, for example) you can decrease the rotation time or slow down the speed. **Note**: The RROS allows control over three speeds (the normal *speed*, a *slow-down-speed*, and a *slow-down-2-speed*) to fine-tune movements. The slow-down-speed is used when a robot without wheel encoders is asked to move a specific distance or to turn a specific amount. Both speeds are set as a percentage of maximum, with 100 being the fastest possible speed using the following commands.

rCommand(SetSpeed, 80)
rCommand(SetSlowDownSpeed, 60)

In general, to calibrate turns, you should set the slow-down-speed to some modest speed, then adjust the time to get a 90° turn. Once you get close to the proper movement by setting the rotation time, you should expect to have to make minor adjustments to the slow-down-speed in order to make the turn accurate. This is true because you have more control over the speed than you do the time (this will make more sense when you actually try to fine-tune your robot).

Fine-Tuning Linear Movements

Next you should calibrate your robot's linear motion. If you tell the simulation to rForward 40, for example, it will move a distance equal to the simulated robot's default diameter. Ideally, your real robot should move a distance equal to its diameter when given the same command. You can fine-tune the linear motion using this command.

rCommand(SetMoveTime,10)

Just adjust the value used until the robot moves *approximately* the right distance. This should generally be done *after* adjusting for turns as it is not as important to have accurate forward movements as it is accurate turns. Just as with turns, robots without wheel encoders will automatically make specific rForward movements using the slow-down-speed.

If you have read any of our other books, you know that the vast majority of the time, your robot is only commanded to move forward one pixel at a time or turn one degree at a time. These movement uses the speed parameter (rather than the slow-down-speed). Normally, the speed should be set to something a little larger than the slow-down-speed.

For most situations, you want the main speed to be as fast as possible, but not so fast that the robot moves a significant amount before sensors can be read and the robot's behavior altered. For example, if your robot is following a line on the floor (as described in some of our books) but it is moving so fast that it looses the line before the line sensors can be read, then you would need to reduce the main speed parameter. There are more sophisticated ways of handling this using a TurnStyle parameter. TurnStyles will be addressed in a later chapter.

Additional Fine-Tuning
It is important to realize that the real robot will normally use the speed parameter when executing an rForward 1 or rTurn 1 command. Typically, you should set the value of speed so that the robot moves only a very small amount for either of these commands. Let's look at a general example to see why this is important. Suppose you were programming the robot to follow a line and that the robot reads the line sensors and either moves forward or turns left or right based on the readings.

If the robot moves too far before reading the sensors again, then it can easily lose the line. When your robot is performing an activity of this nature, it is vital to set the speed parameter to an appropriate value. Of course, a smart robot can use alternative measures to allow it to reliably follow a line at a faster pace. **Note**: Line following examples will be discussed later in this book.

Easier Than You Think
All of the above can sound very complicated but it is important to remember that you only have to perform fine-tuning one time (for each of your robots). Once you have experimented and found the appropriate values, just place all the necessary rCommands into your initialization subroutine and forget about them unless you change things about your robot that might alter its movement characteristics. This could include installing new motors, changing the wheel size, etc.

There will be many other fine-tuning options available to you throughout this text. All appropriate commands should be added to your initialization subroutine. When you have everything exactly how you want it, you can copy and paste the subroutine into a program file of its own and save it. This will allow you to easily merge it (see the FILE menu), or #include it in any of your programs.

Including the file is often the best overall solution because it allows you to simply include the initialization file for the robot you are using at the time. This makes it very easy to use the same program with different robots you own, or even with other RobotBASIC compatible robots at a school or club meeting.

Ramping
The RROS will automatically change each motor's speed slowly so that your robot will not have jerky starts and stops. You can make the robot start and stop quicker by increasing the parameter above 1 (which is the default for small DC motors) in the following command.

rCommand(SetMotorRamp, parameter)

This command is available no matter what type of motors are used to power your robot. You should experiment with different parameters to find what works best for your robot. In general, it is preferred to use the largest parameter that does not cause your robot to jerk or rock when starting and stopping.

Chapter 4

Servomotors

In the previous chapter we interfaced small DC motors directly to the RROS chip. For those that might prefer to use continuous rotation servomotors to power your robot, we added appropriate RROS support. Figure 4.1 shows an early version of the prototype robot we used to test the servomotor routines.

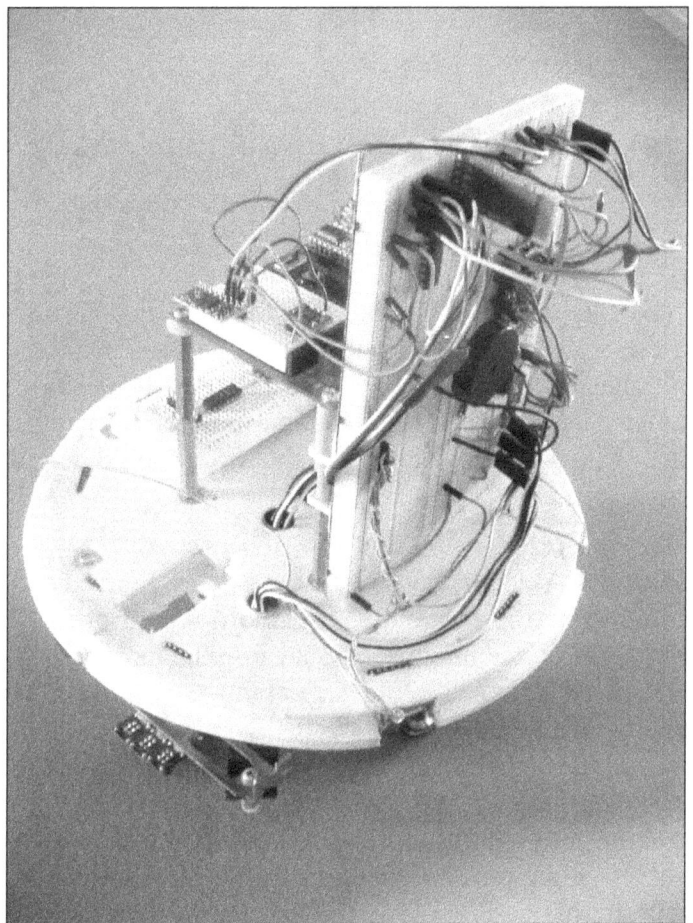

Figure 4.1: This robot was used to test the RROS servomotor routines.

The robot in Figure 4.1 is a heavily modified Boe Bot from Parallax. It was chosen primarily because it uses servomotors and we had it available from a previous project. The Boe Bot's aluminum chassis is small and square, so we topped it with a round piece of foam board to make it look more like our simulated robot. More details on the construction of this project are provided in our book *Robot Programmer's Bonanza*.

Figure 4.2: Servomotors attach to the aluminum Boe Bot
chassis which is topped with a round foam board cutout.

Standard Servomotors
The angular position of the output shaft of standard servomotors is controlled by the width of a pulse sent to them. The pulse width normally ranges from 1ms to 2ms in order to position the output shaft over a 180° range. A pulse width of 1500 microseconds should position the shaft near the center of its travel range. The frequency of the controlling pulse may vary, but in order to achieve smooth movement with a reasonable torque, the servo should be pulsed approximately 50 times per second.

Continuous Rotation Servomotors
Special continuous rotation servos can be purchased from many sources and used to power your robot. The speed of these motors (rather than their position) is controlled by the width of the pulses sent to them. A 1500 microsecond pulse should stop the motor. As the width of the pulse increases the motor will increase its speed in one direction – decreases in the pulse width will increase the motor's speed in the opposite direction.

The RROS will handle all the details associated with servomotor control so that all the commands we used in Chapter 3 to control and initialize DC motors will work equally well with servomotors. There are some additional complications associated with servomotors, so our RROS has a few special commands to fine-tune how they operate. Any new commands needed should be placed in the initialization subroutine just as we did with DC motors.

Servomotors have three connections to them. Often the wires from the servomotor are red, black and yellow. In that case, the black wire is ground, the red wire should be connected to +4.5 to 5 volts and the yellow wire is for the control signal. Sometimes the black wire is brown or grey

26

and the red wire is orange. If your servo has a standard connector, the center wire should always be for power with the darker of the two remaining wires being ground. If you have any doubt which pins are which, refer to the vendor that sold you the device. As with most electronics, improper connections can cause damage.

Most servomotors can operate on 4.5-6 volts without problems, but you should check the specifications for your devices. Our test robot is powered by six rechargeable AA cells in series to get an appropriate voltage for the RROS chip. Since this voltage is too high for most servomotors, you could tap into the supply at 3 or 4 cells to get a reduced voltage for the servomotors, or you could use a 5 volt regulator. This later approach is necessary if your robot is powered from a single 12 volt battery rather than a group of individual cells.

Figure 4.3 shows a 7805, 5V regulator that makes it easy to generate a regulated voltage capable of delivering an amp of current. In order to operate properly, the input voltage to the regulator must generally be at least 7 volts, sometimes higher. If lower voltages are used, the output voltage may only be 4.5 volts or so. Often the 5V devices discussed in this document will work fine at this voltage, but there is no guarantee that erroneous problems will not occur if, for example, you power your robot with a 6V gel-cell battery.

Figure 4.3: Regulators such as this 7805 can produce 5 volts for driving servomotors.

The center terminal of the regulator is ground (connect to the black servomotor wires and to the ground pin on the RROS chip itself). The higher voltage (your 12V battery for example) is applied to the left-hand terminal as pictured in Figure 4.3. The right-hand terminal becomes the 5V source and should be connected to the red leads on your servomotors. This should leave one control [yellow] lead free on each of your servomotors. The control wire for the left motor should connect to the RROS chip, Pin 1. Connect the right motor's control wire to Pin 21. You MUST also connect a resistor (approximately 5K) between each of the above pins and the 5V supply as shown in Figure 4.4.

It is worth mentioning here that companies like Parallax offer large DC motors that can be controlled with pulses just like standard continuous rotation servomotors. We used these motors on our life-sized Arlo robot that is discussed later in this book.

Motor Setup for Servomotors
Of course, we must tell the RROS that we are using servomotors. We can do that with the statement below.

rCommand(MotorSetup, SERVOMOTORS)

You can also use SERVOMOTORS+ENCODERS if encoders are present (see Chapter 7), just as we did with DC motors.

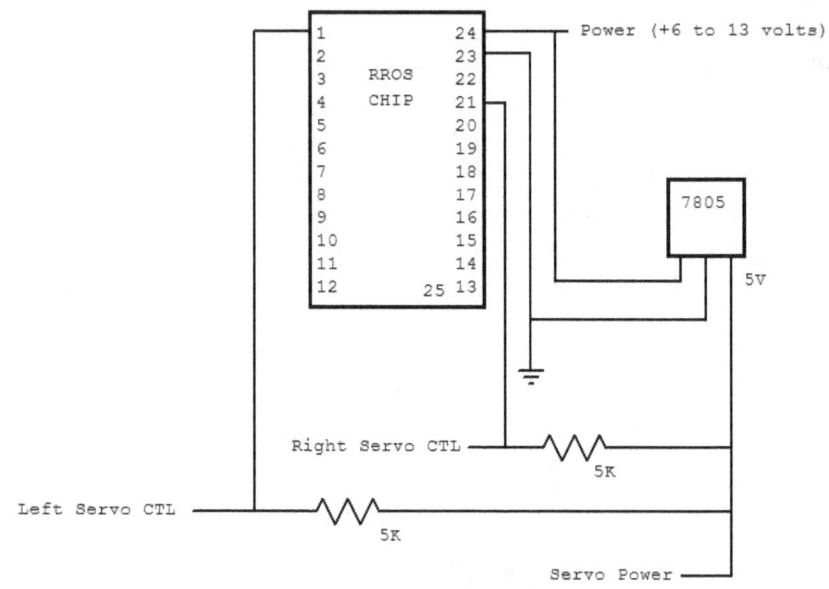

Figure 4.4: The RROS chip can drive servomotors just as easily as DC motors.

You should set the Speed, SlowDownSpeed, RotationTimes, ReducForwLeft, etcetera just as described in Chapter 3 for DC motors, **but before you do so**, you should make sure the control pulses for your servomotors are calibrated properly.

Calibrating the Servomotor's At-Rest State
When your servomotor-powered robot is at rest, the RROS chip will send normally a 1500 microsecond pulse to each motor about 50 times per second. Unfortunately, all servomotors are not exactly alike, and you are likely to find that your servo's will need a slightly different pulse width to make the motors stop moving completely. This is such a common problem with servomotors that the RROS has the option to just quit pulsing a motor when it is suppose to be OFF (this is the default condition). While this does ensure the motors do not turn, even slightly, when they are suppose to be OFF, it is important that we find the true center position for the control pulse if the RROS is to control the motors accurately. This is a very important concept. If the at-rest pulses are not truly causing the servomotors to stop, then there will be a slight bias for one wheel over the other when the robot is asked to move. Calibrating the at-rest pulse can make other commands work better. Let's see how we can calibrate the servomotors.

We can tell the RROS to continue to pulse the servomotors even when they are in an off state by placing the following command in the initialization subroutine. Note: Sending a zero for the second parameter will cause the RROS to return to the normal default state of not pulsing the motors when they supposed to be stopped.

rCommand(CalibServoDrive,1)

This command allows you to calibrate the servomotors that drive your robot. After issuing this command you should expect your robot to drift slightly even when no movement commands are sent to it. In order to correct this situation, watch your robot and determine if either or both of the motors are moving. If they are, you can alter the width of the at-rest pulse being sent to them with these commands.

    rCommand(SetLeftStopOffset,128)
    rCommand(SetRightStopOffset,128)

Notice there is a command for each wheel. The parameter 128 is the default. Making that number larger or smaller (0-255) will alter the corresponding motor's at-rest speed (and eventually the direction of that motor when slowing it down). Once you find the values that make your robot remain stationary (or at least as stationary as possible) when not commanded to move, you can remove the CalibServoDrive command to ensure that the motors remain perfectly still when the robot is at rest.

Calibrating the Servomotors Speed
Most standard servomotors move their output shaft to nearly the same position for similar sized pulses. Of course, the position is never *exactly* the same. The same inconsistency exists with continuous-rotation servomotors, that is, the same pulse width does NOT produce the exactly the same speeds for both motors. Theoretically, for example, a servomotor's movement in one direction should be as described below.

    1500 microseconds   stopped
    1750 microseconds   half speed
    2000 microseconds   full speed

While most servos will probably function somewhat appropriately at the 1500 and 2000 limits, the speed does not usually change linearly. For example, a particular brand of motor might reach 90% of its full speed at 1700 microseconds. If the pulse is increased above 1700, the motor will continue to increase, but the change will be small. In such a case, it would be better for the RROS to assume that the controlling pulse should only vary from 1500 to 1700 so that changes made to the motor's speed parameters (Speed and SlowDownSpeed) will cause reasonable changes to the robot's actual motion. The following command allows you to alter the maximum (and minimum) pulse width used to control your motors.

                    rCommand(SetDriveServoWidth,50)

The parameter 50 represents the default pulse width (50% of normal maximum), and was chosen because it seemed to work best with the motors we tested. You can shorten the pulse with smaller numbers or lengthen it with larger ones. It is not expected that this command will be needed for most motors, but we wanted to provide ways to ensure RROS compatibility regardless of the characteristics of your chosen motors. One nice thing about this command is that it increases or decreases *both* the Speed and the SlowDownSpeed simultaneously (as well as other speed-related parameters to be discussed later in the text).

    Depending on how you mount your motors, they may move in the opposite direction from what you expect (. You can reverse the directions of both of your servomotors using:

                    rCommand(SetDriveServoDir, 0)

A parameter of 1 will return the motor directions to their default condition. **Note:** This command is necessary for servomotors because you cannot reverse their direction by just reverse their leads as you can with DC motors.

After you determine the appropriate calibration parameters for your servomotors, place the proper commands in an initialization subroutine aptly named for your servomotor powered robot so that you can merge it with or #include it in your programs. Utilizing the appropriate initialization subroutine should allow the standard motor commands to work properly whether you are using small DC motors or servomotors. In the next chapter we will examine how to extend this compatibility to much larger DC motors.

Chapter 5

Large Drive Motors

Previous chapters have demonstrated how the RROS can handle all the details associated with small DC or servomotors. In fact, the RROS chip has the hardware built-in to power both of these small motor types. We wanted the RROS to be able to control much larger motors though, when the need arises. We could have built a much larger RROS chip, but that would have added significantly to the price and would not have been needed for many applications. Because of that, we chose to provide the ability to control larger motors by allowing the RROS to interface with any of the RoboClaw motor controllers from BASIC MICRO as pictured in Figure 5.1.

Figure 5.1: RoboClaw controllers allow the RROS to control large motors.

RoboClaw controllers were selected because of their high quality and the variety of products. RoboClaw controllers are currently available that can handle 5, 15, even 30 amps of current for each of two motors. Any of these, or even older models of their controllers, should work find with the RROS. Some models might have slightly different connectors or DIP switch settings, so refer to your RoboClaw documentation to ensure you interface everything properly.

One of the great things about the RoboClaw controllers is that they can be controlled through several modes including analog, RC, and serial. The most efficient method for our RROS is to use is serial so we must set up the RoboClaw's configuration DIP switches and connect everything properly. To make this discussion easier to follow, refer to a drawing of a RoboClaw in Figure 5.2

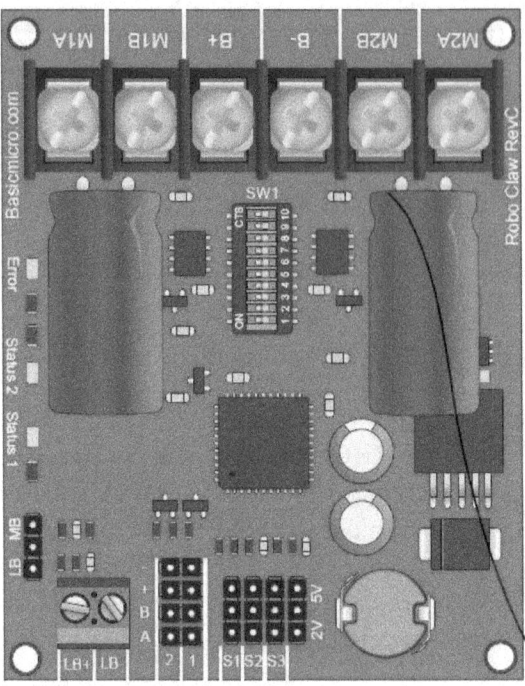

Figure 5.2: This drawing can make interfacing
to the RROS easier to understand.

Setting the DIP Switches

Notice in Figure 5.2, a DIP switch in the upper center of the board. The settings on this switch allow you to configure how the board operates. We need to move switches 2 and 4 to the left, making them ON. This selects **Simple Serial** at **9600** baud. If you are using special batteries (such as Lithium) refer to the RoboClaw documentation as there are additional switch setting to monitor and protect special batteries.

Connecting the Motors

The top of the board (Figure 5.2) has several screw terminals for connecting your motors and the power for them. You should connect the leads from your robots RIGHT motor to the terminals labeled M1A and M1B (upper left corner of the Figure). Your robot's LEFT motor should connect to M2A and M2B. As with small DC motors, if the motors rotate in reverse compared to what is expected, just reverse the leads.

The main power for your motors (probably the battery powering the RROS) should connect to **B+** and **B-** (at the top of the board in Figure 5.2. **Note:** If you use a separate battery for your motors, you must tie the **B-** terminal to the ground terminal on the RROS chip (refer to the RoboClaw documentation).

Connecting the Control Lines

The serial data used to control the RoboClaw must connect to the 3-terminal connector called S1 as shown in the lower middle of Figure 5.2. Only two of the three terminals are actually used for our configuration. The upper terminal (as viewed in Figure 5.2) is the ground terminal and should connect to the RROS chip ground. The lower of the three pins (again, as viewed in Figure 5.2) is the control signal and should be connected to Pin 21 of the RROS chip.

To avoid confusion about this connection, it is worth mentioning that the RROS uses special programming to turn the motor control pins on the RROS chip into a serial port. Because of that, you **MUST** connect a 5K pull-up resistor between the RROS Pin 21 and +5 volts. The exact value is not critical so any values from 4.7 to 10K should work fine.

Motor Setup

As with all other motor options, you must tell the RROS you are using a RoboClaw. This can be done with the following statement.

rCommand(MotorSetup, ROBOCLAW)

As with other motor setups, you can ADD (+) or OR (|) the parameter ENCODERS with ROBOCLAW if encoders are available. Wheel encoders are discussed in Chapter 8.

Calibrating the Motors

The speed and drift of a RoboClaw powered robot should be calibrated the same as small DC motors as described in Chapter 3. Place all appropriate rCommands in an aptly named initialization subroutine so it will be available for future use.

As mentioned in the last chapter, some special DC motors (like those from Parallax, for example) can be controlled by servo pulses. Any servo-compatible motor controller should work with the RROS chip.

Chapter 6

Additional Motion Control Commands

Previous chapters have demonstrated how to connect servomotors and both large and small DC motors to the RROS chip. They have also provided details on how to setup each motor type and perform any necessary calibrations and store them in an initialization file that can be merged or #included into your application programs.

Normal Motion Commands

Typically, a real robot is moved using the same commands as the simulator (rForward and rTurn). We have seen how to fine-tune the RROS so that the real robot responds very much like the simulator. It was also pointed out that the real robot does not need to mimic the simulators actions perfectly because normally an application program will use sensory data to determine how the robot should move. This closed-loop feedback will cause a real robot's reaction to external stimuli to be very similar to that of the simulation. There are some differences though and the RROS has commands that help your real robots act more like the simulation. Chapter 16 provides more insight into this topic.

Turn Styles

The simulated robot has a unique method of turning. It rotates around its center when making a turn by turning one of its "wheels" forward and one backward. This method of turning (compared to a steering system like a car) gives your robot MUCH greater mobility and is usually better for home or office-based robots.

This rotational style of turning works great on the simulator, because the simulated robot does not have any mass or inertia.

A real robot though, does have mass and inertia. Imagine the robot following a line on the floor. When the robot is on the line it simply moves forward. When the robot is slightly off the line though, it would have to turn back toward the line. Since a normal turn for our simulated robot is actually a rotation, it means that one of the robots wheels must actually reverse directions and move part of the robot backward in order to make the turn. This backward motion intermixed with the normal forward motion of the robot will cause a jerky motion. This is even seen on the simulation if you look *very* closely, but it almost unnoticeable.

When it happens on a real robot though, the jerkiness is always annoying and usually unacceptable. Eliminating the jerky motion is actually very easy to fix.

Eliminating Jerky Movement

The easiest way to prevent the jerky motion on a *real* robot is to alter the way it turns. As said, normally the robot executes turns by rotating around its center by turning one wheel forward while turning the other backward. If turns are made by simply stopping one whee though, (instead of reversing it) the jerks are eliminated.

The RROS is capable of handling the alteration of this motion for us. In fact, we gave the RROS even more options when it comes to the way your real robot makes turns. Look at this command.

rCommand(SetTurnStyle, parameter)

The parameter above controls how fast the *slow* motor moves when the robot makes a turn. When making a left turn, for example, the left motor should be slower than the right motor, or perhaps even stopped as discussed above.

If the parameter is set to zero, then the slow motor will indeed be stopped when the robots turns. This lets the motor effectively rotate around its slow wheel rather than rotate around its center. If the parameter is set to 50 though, the RROS will reduce the speed of the slow wheel down to 50% of that of the fast wheel letting the robot make a very slow, loopy style of turn. A parameter of 90 will not really make the robot turn much at all. Since the slow motor will be moving at 90% of the speed of the faster motor, the robot's motion will be more of slight drift away from a straight line movement. **Note:** It is because of commands like this that servomotors have to be calibrated so that their speed changes more linearly over the entire usable range (refer to the rCommand SetDriveServoWidth in Chapter 4).

A parameter of 100 returns the robot's behavior to the rotational style of turning. It is worth mentioning, that the RROS handles turning styles very intelligently. If you ask the robot to turn 45°, for example, it will *always* use a rotational style. When the rTurn argument is 1 or -1 though, the previously selected turn style will be used. As you develop behaviors, you will see that this is exactly how you would want the robot to behave.

For most types of sensor-controlled movements (following a line or hugging a wall, for example) setting the TurnStyle to a low number (0-50) produces an acceptable non-jerky movement. An intelligent robot though, can modify the parameter on-the-fly and greatly improve the performance for a given task.

For example, if the robot is following a line that is only moderately curved, it can do so much quicker when a slower turning style is used (perhaps 60 or 70). If the line is curvy though, a turn style of 10 or 20 might be more appropriate. As shown in some of our other books, it is not difficult to use sensory data to determine how curvy a line is (the more *often* a robot has to turn, the curvier the line). An intelligent robot should keep track of the line's curviness, and constantly adjust the turn style parameter to ensure the robot's actions are always at peak performance.

Turning to Specific Angles

Once you start programming your robot to perform meaningful tasks, you may find the need to have it face a given direction – such as North. Because of that, we have provide a command that tells the robot to move to a specified angle (assuming your robot has a compass). You could, of course, perform this action using only RobotBASIC commands, but we think you will find this special feature fast, efficient, and very useful. Complete details on how to use this movement command will be covered in Chapter 10 where the RROS compass is discussed.

RROS Timeout

Normally when the simulator is give the command rForward 1, it moves forward on the screen by a single pixel. If the real robot tried to move such a small distance and stop until it received another command, it would exhibit a jerky movement. This happens because Bluetooth (and most other wireless links) must delay each time they switch from transmitting to receiving. Because of this delay, the RROS (when using a Bluetooth connection) can only process about 10 to 15 commands each second which means the robot could set idle for 100ms while waiting for the next command.

To prevent this, when rForward commands of 1 or -1 are used, the RROS will normally keep the motors moving at the designated Speed until a new command is received telling it to do otherwise. This action is handled automatically by the RROS and does not require any action on your part. Keeping the motors moving has the potential to be a problem though, unless the RROS is smart enough to turn off the motors when no command is received within a specific amount of time. Look at the following program fragment which starts the motors and then ends – which means the RROS will not receive another command. If the RROS simply kept the motors running in situations like this, the robot would continue to move at its last known speed even after the program ended.

```
rForward 1
end
```

To prevent this from happening, the RROS has a designated default time-out period. If no command is received in that period the RROS automatically stops all drive motors. You can use the command below to set a different time-out period based on your needs.

```
rCommand(SetRROStimeout,15)
```

If the time-out period is too low, the robot will have a jerky motion because the motors are actually turning on and off while the robot moves. If the time-out is too long the robot may continue moving too much under circumstances like that described above. Generally you should not have to alter the time-out period because if your program is doing something that might take considerable time, such as waiting for a voice command, it is probably advisable to stop the robot's movements completely. If your RobotBASIC application is doing many calculations before sending the next rForward command though, you may want to extend the timeout period slightly to prevent the inevitable jerky motion.

This is just another small example of how the RROS was designed so it could be customized to meet your needs.

Wired Links

Remember, speed related problems are usually caused by the wireless link itself. The RROS itself is actually capable of processing commands somewhat faster. For high-performance situations, we suggest that you embed a laptop or netbook running RobotBASIC inside your robot and connect to the RROS chip using a *wired* serial connection. This will improve the speed communication and let you obtain and react to sensory data quicker than a wireless link. In most situations though, a wireless link is fast enough as demonstrated by our life-sized Arlo robot (discussed later in the text).

More Commands

There are even a few more commands that alter the way your robot moves, but they deal specifically with wheel encoders, and will be covered in the Chapter 8.

Chapter 7

Steerable Robots

In previous chapters, we examined how the RROS could control a wide variety of motors in a manner similar to that of the RobotBASIC simulated robot. As you know, the simulator turns left and right by independently controlling the speed and direction of the two drive wheels. The robot rotates around its center, for example, when one wheel moves forward with the other in reverse – an action very different from a steerable vehicle such as a car.

The rotational steering of the simulator allows it to maneuver in a cluttered environment far easier than a steerable robot and should probably be the choice for any home or office-based robot. When robots must operate in a more expansive environment such as open field though, the steerable robot can be more stable. Note: Another option is to use a four-wheel (for stability), all-wheel-drive (one motor on each wheel), robot that still turns like the simulator. Simply place the two motors on each side in parallel.

MINDS-i

A company called MINDS-i (mymindsi.com) sells a wide variety of steerable, all-terrain kits and construction sets. Figure 7.1 shows a sample robot from their web page. Notice the use of knobby tires and a spring and shock absorber suspension designed to improve performance in a rugged landscape.

The RROS Can Steer

Even though RobotBASIC was not designed to simulate an all-terrain vehicle, we wanted to make the RROS capable of handling a four-wheeled robot where one motor drives the rear wheels in unison and another motor steers the front wheels. In order to invoke this mode, simply tell the RROS you are using a steerable system as shown below.

rCommand(MotorSetup, Steerable)

In this mode, the rForward command will produce appropriate signals on Pin 1 of the RROS chip to control a continuous rotation servomotor for driving a robot's rear wheels. Since all-terrain vehicles will generally require larger motors, MINDS-i powers their robots with a DC motor controlled called the 300A Electronic Speed Controller or 300A ESC for short. This device, allows a single DC motor to be controlled using RC (servomotor) pulses.

In this mode a standard servomotor is assumed to be used to move the robots steering mechanism, so the rTurn command will produce servomotor pulses on Pin 22 of the RROS chip. Note: Pull-up resistors must be used on both Pins 1 and 22 as described in Chapter 4. When properly calibrated (more on this later) an rTurn 15 will make the servomotor turn the front wheels

15° to the right while an rTurn -30 will turn the wheels 30° to the left. These actions make it easy to control a real robot, but it would be nice to be able to simulate a steerable robot within the RobotBASIC environment.

Figure 7.1: Steerable robots, such as this all-terrain vehicle from MINDS-i offer increased stability in an outdoor environment.

Simulating a Steerable Robot

Even though the RobotBASIC simulator only deals with two-wheeled robots, we can simulate a steerable vehicle with a little trickery as shown in Figure 7.2.

```
Main:
 // draw a line for reference
 line 250,300,700,300,3,green
 // the following line replaces rLocate
 call InitSteerBot(230,300,90)
 call Steer(0) // replaces rTurn
 // note: steer works in 10 degree increments
 call Drive(10) // replaces rForward
 for i=1 to 130
  call Drive(1)
  if i=25 then call Steer(-10)
  if i=50 then call Steer(0)
  if i=75 then call Steer(30)
  if i=100 then call Steer(-20)
```

```
  delay 50 // to reduce robot's speed
 next
end

sub Steer(amount)
 rGPS x,y
 CD=rCompass()
 for i=CD+MI[0] to CD+amount
  t=DtoR(i-90) //(CD+MI[0]-90)
  rLocate x,y,CD
  line x,y,x+13*cos(t),y+13*sin(t),2,RED
 next
 MI[0]=amount // wheel direction
return

sub Drive(Amount)
 for i=1 to Amount
  CD=rCompass()+MI[0]/10
  rGPS x,y
  restorescr
  //CD+=delta
  rLocate x,y,CD // face wheel dir
  rInvisible Green
  if MI[0]=0
   rForward 3
  else
   rForward 5
  endif
  //draw wheel
  rGPS x,y
  t=DtoR(CD+MI[0]-90)
  line x,y,x+13*cos(t),y+13*sin(t),2,RED
 next
return

sub InitSteerBot(x,y,d)
 SaveScr // preserves environment
 // create a single element global array
 data MI;0
 rLocate x,y,d
 MI[0]=d-90
 t=DtoR(MI[0])
 line x,y,x+13*cos(t),y+13*sin(t),2,RED
return
```

Figure 7.2: This example program demonstrates
how to simulate a steerable robot.

The program in Figure 7.2 is provided for illustration as it has some limitations. Commands such as rSpeed and rPen, for example, do not work properly and the robot moves numerous pixels for an rForward 1 command. All the sensor commands work properly though, allowing the subroutines to be truly helpful when developing behaviors and prototyping algorithms for a steerable robot. **Note:** It is even possible to utilize the standard simulator to help develop algorithms for a steerable robot. Refer to Chapter 16 for more on this.

If you create some applications with these routines you will see that steerable robots do not operate well unless their perimeter sensors have a significant range. This is true because they must make turning decisions far sooner than a two-wheeled robot. This means that the RROS's Virtual Sensors discussed in Chapter 9 are an ideal solution for an all-terrain vehicle.

Since all-terrain robots generally work in a rugged, demanding environment, it is often worth the expense of a commercial kit or construction set that offers improved suspension options. If you just want to experiment before committing to a large investment though, there are low-cost alternatives.

Modifying Toys

When developing our steerable RROS routines we needed a low-cost prototype and purchased the toy truck shown in Figure 7.3 from Radio Shack.

Figure 7.3: This radio controlled toy truck served as the basis for our steerable prototype.

Toys such as the truck we used are built for speed and would be very hard to control as robots. Also, the steering system on most low-cost toys only have three states – full left, straight, and full right. Therefore, our first step was to disassemble the unit and discard all the original electronics, including the radio receiver, and both the drive and steering motors.

We replaced the original drive motor with a small heavily geared motor as shown in Figure 7.4.

Figure 7.4: A small, heavily geared motor was mated to one of the original drive gears.

The new motor must be precisely positioned or the gears will either bind or slip. It might have be easier to utilize a continuous rotation servomotor for our small prototype, but we needed a DC drive motor to ensure our compatibility with the controller used in the MINDS-i robots. To that end, we interfaced our motor with the Electronic Speed Controller shown in Figure 7.5. The ESC has two independent wires to connect to the motor and a battery connector often used on RC cars. These will have to be cut off unless you are using standard devices. Before you do any cutting though, mark each wire to ensure you do not confuse them later. The ESC also supports an ON/OFF switch and a standard servomotor connector which we will use to interface with the RROS.

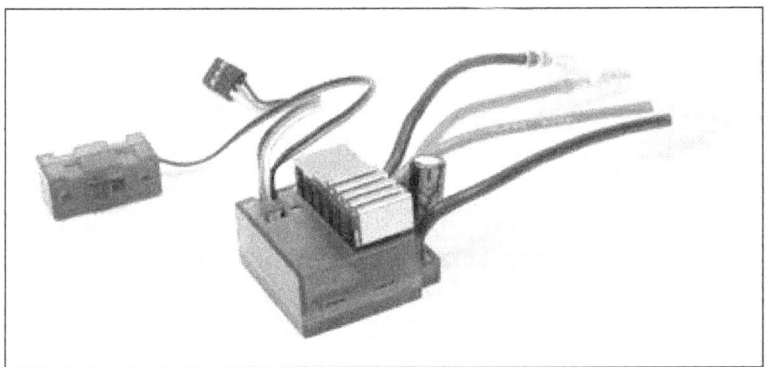

Figure 7.5: This 300A ESC is used to control

the drive motor in MINDS-i robots.

MINDS-i robots utilize a servomotor to control their steering, so we did the same. Our toy truck had a great steering mechanism that was easily interfaced with a micro servo as shown in Figure 7.6.

Figure 7.6: The servo horn moves a rack-bar connected to the wheel assemblies.

Calibrating the RROS Commands

As with all our RROS commands, we must have ways of calibrating them to ensure compatibility with a wide selection of hardware options. This is especially true for the steering command, rTurn.

As mentioned earlier, a command of rTurn 15 should turn the robot's wheels 15° to the right. When you issue this command, your robot's wheels should turn, but without calibration they probably will turn too much or too little, and, depending on how you mounted your steering servo, they might turn left instead of right.

We can adjust the total amount of steer turn (which is controlled by the width of the servomotor pulse) by issuing rCommand(SetSteerServoWidth, parameter) where a larger parameter will move the motor further for any given rTurn value. Use small parameters such as 25 to start and increase them slowly until you get the desired action. Large numbers can move the servo past its limitations, possibly damaging the gears. The steering direction can be controlled using rCommand(SetSteerServoDir, parameter) using parameters of 1 or 0.

The maximum pulse width for the drive motor can be controlled with rCommand(SetDriveServoWidth, parameter) just as we used in Chapter 4. The direction of the drive motor can be reversed with rCommand (SetServoDir, parameter) which was also discussed in Chapter 4.

Our Completed Steering Prototype

Figure 7.7 shows our completed prototype. It uses six rechargeable AA cells, four of which are in the trucks original battery compartment. A small breadboard makes it easy to connect the RROS chip and other electronics. Because our prototype was needed only to test the new motor commands, we did not permanently add any sensors to this robot. Since an increased sensor range is needed (as mentioned earlier) we highly recommend our Virtual Sensor System discussed in Chapter 9. As an alternative, MINDS-i offers a digital sensor with a manually adjustable range that provides a slightly lower cost, but with far less flexibility.

Figure 7.7: Our steering robot prototype.

Chapter 8

Wheel Encoders

In Chapter 3, we discussed how you could specify the timing periods your robot should use when it moves forward or turns a specified amount. We also examined how to fine-tune the speed of each wheel so that the robot could track forward and backward in a straight line. All of these methods can produce acceptable results, especially if your robot always has a fully charged battery and is operating on a level surface.

The Need for Encoders

Sometimes though, your robot might need more accurate movements. One way to accomplish this is through wheel encoders. Some motors can be purchased with an integrated encoder, and there are commercial versions of course that can be added to compatible motors. You also have the option to build your own encoders.

The principle for an encoder is simple. You simply need a device that can generate a string of pulses as your robot's wheels move. Our RROS can constantly count these pulses and keep track of how far the robot has moved or turned.

Simplified Approach

The approached used by our RROS is very basic, yet it provides excellent functionality. We considered using the more robust *quadrature* encoding but decided against it for two reasons. First, quadrature encoders require two detectors on each wheel that must be precisely aligned. Second, we found ways of making simpler technology meet our needs. Since our system is composed of only a single detector on each wheel, it requires that the RROS do more work, but that, after all, is its job.

Types of Detectors

The RROS does not care what kind of hardware is used to create the pulses indicating your robot's wheels are moving, so advanced hobbyists can consider any option. You could, for example, use a standard quadrature encoder if you have one available or if one is built into the motors you are using – just use *one* of the encoder's outputs and leave the other disconnected.

If you want to build your own, a simple reflective sensor is the easiest to implement. Basically, an LED in the sensor emits a small amount of light (usually infrared) and a phototransistor detects if that light is reflected back or not. This means that your robot's wheels (or even one of the gears in the train) has to be equipped with some form of encoding disk so that pulses are generated as the disk spins by the sensor. Figure 8.1 shows a representative disk that can be glued to a wheel (refer back to Figure 3.2 in Chapter 3).

Figure 8.1: This disk will produce 18 counts for each revolution of the wheel.

Notice that the disk in Figure 8.1 has 18 spokes, 9 white and 9 black. This will produce 18 pulses each time the disk makes one complete revolution. The more spokes you have, the more accurately the RROS will be able to track your robot's movements. The number (and thus the size) of the spokes are limited though, by the size of the sensing element. If, for example the lens on your sensing element is large enough to cover multiple spokes, then it will not be able to generate pulses correctly. Commercial encoders will generally provide much better performance. If you use a commercial quadrature encoder, just use ONE of the two available outputs (either will work).

The RobotBASIC program in Figure 8.2 was used to draw the disk in Figure 8.1. It allows you to specify the size of the disk as well as the number of spokes.

```
// alter these parameters for your disk
r = 100 // radius
n = 18  // number of spokes
x = 400 // center of disk
y = 300
LinesPerSpoke = 360/n
LineWidth 1+r/50
c=Black
for i=0 to 360
 if i#LinesPerSpoke = 0
  if c=Black
   c=white
  else
   c=Black
  endif
  SetColor c
 endif
 line x,y,x+r*cos(DtoR(i)),y+r*sin(DtoR(i))
next
SetColor Black
LineWidth 2
arc  x-r-2,y-r-2,x+r+2,y+r+2,0,6.28
circle x-3,y-3,x+3,y+3
```
Figure 8.2: This program can draw encoding disks for you.

The program in Figure 8.2 uses wide lines to ensure that the black areas are totally filled in. This causes some minor distortion at the center of the disk. This should not be a problem because the reflective sensor should be mounted as close to the outside edge of the disk as possible (where you have the widest possible spokes). **Note**: This program requires that the number of spokes be a number that can divide evenly into 360, otherwise one of the spokes will differ in size from all the others.

The Sensing Element

As stated earlier, you can use any sensing element you wish. Pololu offers the QTR-1A which is very inexpensive and pictured in Figure 8.3. **Note**: Pololu also offers a QTR-1RC that is designed to be used in an entirely different way. It is NOT compatible with our RROS.

Figure 8.3: The QTR-1A from Pololu is an inexpensive reflective sensor.

The large hole in the QTR-1A is used for mounting. We used a screw to attach the sensor to a small block of wood, then glued the block to the robot's body so that the sensor rested about $\frac{1}{8}$ of an inch above the surface of the encoding disk. If the disk is too close or too far away, the sensor will not operate properly. Ideally you would monitor the sensor output using an oscilloscope while mounting it. This sensor has a reasonable amount of tolerance, though, so even monitoring the change in the output voltage with a meter as you slowly rotate the wheel should allow you to calibrate its mounting position properly.

It is also very important that the sensor be parallel to the disk. If it is tilted, the light generated by the sensor will be reflected off at an angle and never be returned to the receiving element. Mounting the sensor sounds more complicated that it actually is. Just be happy you do not have to mount and align two sensors together on each wheel (which would have been required for quadrature encoding).

There are three holes at one end of the sensor (see Figure 8.3) that provide the connections to the RROS circuit. You may solder wires directly to the holes, or solder pins as shown in Figure 8.4 so you can connect to the sensor to your circuits with a standard servomotor cable.

Figure 8.4: Soldering pins to the connector can make interfacing easier.

Figure 8.5 shows a schematic of the actual sensor circuit. The two main elements are an IR LED to produce the light, and a phototransistor for detecting the reflected light. Vin should be connected to a 5 volt source on your robot (refer to Figure 4.3, Chapter 4).

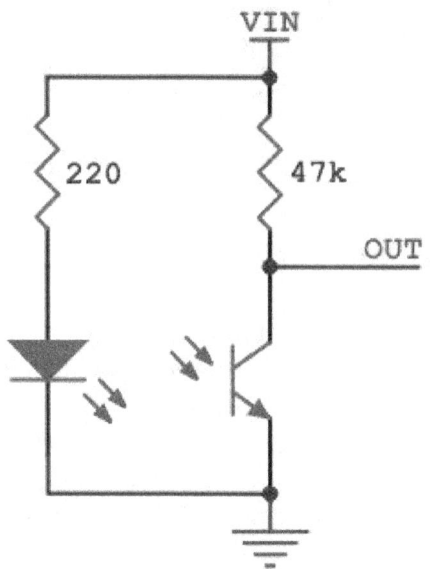

Figure 8.5: This is the actual schematic of a QTR-1A.

When the photo transistor in Figure 8.5 receives light, it conducts and forces the output to zero. When no light is reflected, the output is high (5V).

Connecting to the RROS Chip

The output from the encoder on the left wheel connects to Pin 3 on the RROS chip. The right encoder output connects to Pin 4. When you issue a motor setup command, you can OR or ADD the parameter ENCODERS as described in Chapter 3 to tell the RROS to start counting the pulses. You can also use rCommand(EnableCounters, parameter) with either a TRUE or FALSE (or 1 or 0) parameter to turn the encoders ON and OFF during normal program operation.

In most situations, the signal from a circuit such as the one in Figure 8.5 will work just fine, especially if the sensor is placed properly in relationship to the encoding disk. If you find the pulses generated by the sensor are not *clean* enough, you can run them through a standard inverter (such as a 7404 or 74HCT04) or better still a Schmitt Trigger inverter (such as a 7414 or 74HCT14) before they are connected to the RROS chip.

Using the Encoders

The RROS will handle operations that utilize the encoders automatically. When you tell the robot to rTurn 90, for example, it will count the pulses and stop the turn at the right position. It will also ramp your robot up to speed and slow it down incrementally to prevent jerky starts and stops. In order for it to do its job though the RROS must be initialized with the appropriate information using the following commands.

```
rCommand(SetClicksPerDiam, Param)
rCommand(SetClicksPer90, Param)
rCommand(SetMotorRamp, Param)
```

In the first statement, Param should be the number of clicks your encoder counts when your robot moves a distance equal to its diameter. The easiest way to find this number is through experimentation. Write a program that tells the robot to move rForward 40 and then ends. Start with a value around 30 in the first statement above, and if your robot is moving too far, reduce it. If the robot does not move far enough, increase it. Do the same thing using an rTurn 90. If you have trouble performing this calibration, try adjusting using lower values for the standard speed and the slow-down-speed.

The default Param for the motor ramp is 1. As the numbers get higher, the robot will start and stop more quickly. You should test different numbers and utilize the largest ramp value that does not cause your robot to start and stop too abruptly. Using ramp values that are too low will limit how well the robot perform in some situations. Examples are given in Chapter 16.

Advanced Operations

The RROS will try to help you in other ways when the counters are enabled. For example, if you ask the robot to move with an rForward 100 command the robot will ramp up to the normal Speed, and move until it gets close to its specified destination. At that point the RROS will slow the robot down to the SlowDownSpeed and continue until the end point is reached. It is important to realize that the RROS does not have to be told to do this. It is just part of its job if counters are enabled. You will find that the RROS's actions will not only help your robot move smoothly, but also make it look more intelligent.

You can also specify a SlowDown2 speed. If set to anything other than zero, the RROS will automatically reduce the speed again (to this value) when the robot is *very* close to its destination. This effectively creates a proportional control that allows the robot to move much faster than when open-loop control is used.

Here is another way the RROS will provide help without needing any input from you. Anytime your robot is trying to do a movement where the left and right motors should be turning at the same speed (example: normal turns or moving in a straight line) the RROS will monitor the counts for each wheel and automatically slow down the faster wheel until the two counts are equal again.

The speed of the slow wheel will be reduced by its normal speed divided by a Count-Correction-Divisor. If for example, the Count-Correction-Divisor is 2, the speed of the slow wheel will be

51

reduced by ½ of its normal speed. If the divisor is 5, the speed will be reduced by 20%. The default divisor is 2 but it can be set by changing Param in the following statement.

rCommand(SetCCdivisor, Param)

Commercial Encoders

Home-made encoders like the one described in this chapter are generally limited by the optical characteristics of the detector used to count pulses. This is not really a problem because the RROS can do a *reasonable* job even if the robot only generates 30 pulses or so when it moves a length equal to its diameter. It does a better job though, if the encoders can produce three or even four times that number of pulses.

Some commercial encoders though can produce far more pulses per revolution than the RROS needs or perhaps can even handle. After all, the RROS is doing a lot of actions in the background. If you are using a commercial encoder capable of producing more than 255 pulses as the robot moves a distance equal to its diameter, you have a problem because when you use SetClicksPerDiam, the Param can only be an 8-bit number (having a maximum value of 255). In such situations, you can have the RROS automatically divide the counts it accumulates by some Param with the following statement.

rCommand(SetCountDivisor, Param)

If the Param is 2, for example, only half of the encoder pulses will be counted. Normally you should never have to utilize this command, but it has been included to handle improbable situations. In the rare case that you use an encoder that produces pulses faster than the RROS is capable of counting, you will have to reduce their frequency by using an external counter chip.

Fine-Tune First

When wheel encoders are used, the RROS will automatically correct for many situations as described in this chapter. The RROS may be able to do its job more efficiently though, if the speed of the drive motors are *already* reasonably matched. For that reason, if you have trouble you should try initially turning off the counters and fine-tuning the motors using the techniques discussed in Chapter 3. This will ensure that whatever motors you are using are as closely matched as possible, thus minimizing the corrections needed by the RROS.

Final Thoughts

Wheel encoders certainly can help your robot to turn and move more accurately, but do not assume perfection. Gears still have slop and backlash and tires still have friction and slip so there will still be error in the robot's movement even when the wheels move the appropriate amount.

Sit your robot up on blocks and place a chalk mark on both wheels, then tell the robot to rForward 100, then rForward –100. You will see that the wheels return to their original position if encoders are enabled. If you do the same program with the robot on the ground, it should move a relatively straight line and return to somewhere near its starting point. If you see it drifting left or right, make sure the wheels are aligned properly as weight distribution can significantly affect movement when the wheels are misaligned. If you continue to have drift, try giving each wheels a slight toe-in (angle each wheel slightly to the center).

Remember, your robot should seldom move any significant distance without using its sensors to guide its actions. Because of that, the wheel encoders need only to provide reasonably accurate movements.

Chapter 9

Perimeter Sensors
Turret Ranger & Battery Monitoring

This is one of the most complicated chapters in the book - not because the subject itself is complicated, but because there are so many options available to you. When it comes to perimeter sensor types, the shear number of choices the RROS gives you can be a bit intimidating, but the large number of possibilities gives you enormous flexibility when it comes to designing your robot. You will see that the choices you make when it comes to perimeter sensors will also affect the turret ranger and battery monitoring, which is why they are included in this chapter. We will do our best to present this material so that it is as easy to use as possible.

Normally, a RobotBASIC robot will have two basic types of perimeter sensors. Figure 9.1 shows the placement of four bumper sensors which can be read using the rBumper() function and the placement of five proximity sensors that can be read with rFeel().

Bumper sensors only detect objects very close to the robot and could actually be implemented with physical bumpers that detect collisions. The proximity sensors are assumed to be IR or ultrasonic sensors that can detect objects before any collision has occurred. As depicted in the figure, proximity sensors generally detect objects within a cone-shaped area outward from the robot. When IR sensors are used, the width of the cone can be very narrow, sometimes even approximating a straight line. Certain brands of ultrasonic sensors (such as Maxbotics) allow you to choose the shape of the detection cone by purchasing specific models. Choose your shape based your application.

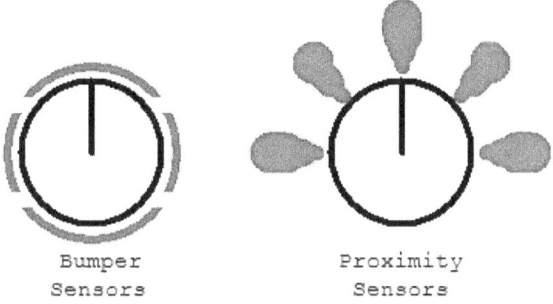

Figure 9.1: RobotBASIC robots have two types of perimeter sensors.

For many behaviors, the proximity sensors provide all the capability you will need. For advanced behaviors though, it is nice to have both options available to you – letting you know when objects are close (proximity), and when they are *really* close (bumper).

Categorizing the Possibilities

In order to make it easier to visualize all of the choices available to you, let's organize them into categories based on the type of perimeter sensors used. This makes sense, because all other sensor types (compass, beacon detector, line sensors, etc.) are always available no matter what type of perimeter sensors are used.

First, let's divide the choices into two major options – *digital* perimeter sensors or *ranging* perimeter sensors.

The Digital Perimeter Sensor Mode

Digital sensors are generally cheaper than ranging sensors so this mode usually allows you to make the cheapest robot possible. This mode does not have bumper sensors though (unless a sensor expansion chip is used – more on this custom option in a later chapter). This means your robot's perimeter sensing will be handled by five digital proximity sensors mounted as depicted in Figure 9.1. As mentioned earlier, the proximity sensors are sufficient for many beginner-level projects. In fact, if the user is totally new to programming then having only one type of perimeter sensor available can often be less confusing.

With the RROS, you can use any sensor you wish in this mode as long as it provides a LOW signal when an object is detected and a HIGH signal otherwise. The Sharp GP2Y0D810 IR sensor (Pololu #1143, shown in Figure 9.2) is an excellent choice. It can detect objects up to four inches away. For very small robots, there is also a version of this sensor with a detection range of two inches.

Figure 9.2: This Sharp digital IR sensor makes it easy to detect objects close to your robot.

There are only three connecting pins on the Sharp sensor. Two are for connecting 5V and ground, and the third is the output that will be connected to the RROS chip (more on this later in the chapter).

If you utilize digital perimeter sensors and want to be able to use the rRange() function (which returns the distance to objects at a specified angle relative to the robot's heading) then you will have to build a servomotor-controlled rotating turret on which you mount a ranging sensor. We mounted a miniature servomotor in a foam-board stand as shown in Figure 9.3. The chosen ranging sensor can then be hot-glued to the servo's horn. You may use any of the supported ranging sensors.

Figure 9.3: Raising the turret motor helps ensure other components will not block the ranging sensor.

We will see later that a turret mounted ranger will only be necessary when the digital mode is used. When needed, the servomotor for the turret must have 5V and ground connected to it just as described in Chapter 4 for the servomotors used to drive the robot. The servo's control signal pin should connect to Pin 10 on the RROS chip.

The Ranging Perimeter Sensor Mode

In general, this option *usually* uses six ranging sensors (either IR or ultrasonic) spaced around the robot as shown in Figure 9.4.

Since each of the sensors shown in Figure 9.4 is a ranging sensor, it can *measure the distance* to objects in its path. The RROS will use this capability to create a *virtual sensor system* (VSS) that implements the bumper and proximity sensors.

You will be able to specify two distance parameters with special rCommands. We will call these two distances BumpDist and ProxDist. Let's see how these levels are used.

55

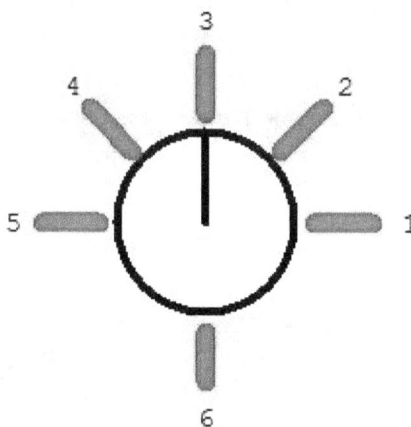

Figure 9.4: Ranging sensors placed as shown can produce a virtual sensor system.

A Virtual Sensor System

When the RROS detects objects closer than BumpDist it will automatically react as if a bumper sensor at that position was activated. Objects detected closer than ProxDist will be seen as a proximity sensor. This is a very powerful concept that allows our robot to *effectively* have both bumpers and proximity sensors even though they do not exist in that form.

When ranging sensor 1 detects objects within the BumpDist it will show up as the right bumper. In a like manner, ranging sensor 5 translates to the left bumper and ranging sensor 6 translates to the back bumper. The front bumper will appear active if ranging sensor 2, 3, *or* 4 detects an object within BumpDist. We are not done yet, though.

A Virtual Turret

The simulated robot has a (simulated) ranging sensor mounted on a rotating turret allowing it to measure distances to objects at any specified angle and report that distance using rRange(). examine an area defined by a 180° arc in front of the robot and provide the distance to objects found using the function rRange(). As mentioned earlier, the RROS actually supports using a real turret in the Digital Perimeter Sensor Mode mentioned earlier. **Note:** The real digital turret is generally confined to view an area of ± 90° from the robot's heading due to limitations of the servo itself.

In the Ranging Perimeter Sensor Mode, the RROS will simply report the distance recorded by the range sensor closest to the desired turret angle requested. This adds even more power to the VSS concept because the area in front of the robot can be scanned without the expense of having to build a turret. The rRange() operation will also be faster, because there is no waiting for a turret to move. Note: As a special added feature, if the command rRange(1) is used, it will return the distance measured by the rear sensor. This may seem strange, but is necessary because RobotBASIC restricts rRange() commands with the real robot to ± 90 and since a request of 1◻ would normally revert to 0◻ anyway, this causes no conflict.

In summary, the VSS lets six ranging sensors act as four bumpers, five perimeter, and a turret mount ranging sensor.

Supported Ranging Sensors

The RROS currently lets you choose from five different ranging sensors (or their equivalents) as listed below. Two are infrared sensors and three are ultrasonic.

- Sharp GP2Y0A21 short range IR (Pololu #136)
- Sharp GP2Y0A02 long range IR (Pololu #1137)
- Maxbotics LV-MaxSonar-EZ family (Pololu #723 and others)
- Parallax Ping))) Ultrasonic (Pololu #1605)
- SR04 Sonic Rangers (available at RobotBASIC.com)

The Pololu item numbers are listed but these items are generally available from other vendors such as Parallax, Lynxmotion, The RobotShop, RobotMarketPlace, TrossenRobotics, JameCo, Digi-Key, and many others.

Comparing IR and Ultrasonic Sensors

We feel IR sensors are generally better for small robots because emissions from one ultrasonic sensor can be read by another, thus providing unreliable data. This can be minimized by using directional hoods made from rubber tubing (see Appendix D). The RROS reads the PING sensors in two groups of alternate sensors to minimize this problem. Maxbotics and SR04 sensors must all be read simultaneously though, so directional shields must be used to prevent false readings from alternate echo signals on small robots.

IR sensors are also not without their problems, as small objects located between the narrow IR beams can be missed. The wider detection cone of ultrasonic sensors is a good reason to choose them for larger robots.

There is no right answer when it comes to sensors. IR can only detect objects that will reflect infrared light, and ultrasonic sensors can have trouble with soft objects that cannot reflect sound as well as objects with angular faces that bounce the sound waves away from robot rather than back to the sensor. Choosing the correct sensor for *your* application is very important. We feel that the needs of most applications can be met by one of the supported sensors. We wanted the RROS to be able to handle almost anything though, so Chapter 13 will discuss methods of using several types of sensors in unison, and even advanced sensor options such as vision.

Even More Options

Both the IR and ultrasonic sensor configurations have more options. Instead of using six ranging sensors as described above, you can use a digital sensor on the backside of your robot letting you reduce costs as long as you do not need to vary the detection distance on the rear of your robot.

Choosing the Sensor Mode

Just as we used an rCommand to perform the motor setup back in Chapter 3, the following command can be used to initialize the sensors. **NOTE:** Setup motors before sensors .

rCommand(SensorSetup, Param)

The value of Param will tell the RROS what sensors you are using. The two most significant bits indicate the main mode as shown below.

00 All Digital
01 5 ranging sensors (rear digital)
10 6 ranging sensors

The lower 3 bits of Param indicate the type of ranging sensor that is being used. In the Digital Mode, these bits indicate the type of ranging sensor used on the turret. These options refer to the supported sensors listed earlier and are indicated by the numbers given below.

000 Short IR (Sharp GP2Y0A21)
001 Long IR   (Sharp GP2Y0A02)

> 010 Analog Sonar (Maxbotics)
> 011 Ping   (Parallax)
> 100 Sonic ranger (RobotBASIC.com)
> 111   No ranging sensor available

Just as with the motor setup, you can use the actual numeric values for Param as described above, or you can use the variables predefined in RROScommands.com. For example, you can set up a robot with digital perimeter sensors and a turret mounted Ping sensor with this statement.

rCommand(SensorSetup, DIGITAL+PING)

Similarly, the following command tells the RROS you are using six Maxbotics sensors.

rCommand(SensorSetup, SIXRANGE+MAXBOTICS)

This SensorSetup command can be used to set up the compass too, but that will be discussed in Chapter 10.

General Wiring for the Perimeter Sensors

Since the RROS supports numerous types of perimeter sensors, we will first examine the general connections that must be made for each of the sensor types before discussing how the sensors connect to the RROS chip itself. Usually these general connections are only for power and ground, but there are exceptions.

The Sharp IR Ranging Sensors

Figure 9.5 shows a Sharp GP2Y0A21 which can detect objects up to about 30 inches. Its longer-range cousin, the GP2Y0A02, looks very similar, but is slightly larger. Both devices have 3 connecting terminals, two of which supply 5V and ground. The third terminal for each sensor will connect to the RROS chip as described later in this chapter. The RROS will read the analog voltage on this pin and convert it to a distance reading. The readings for the Sharp IR sensors are very fast, but its analog readings are nonlinear so they are not as accurate as the ultrasonic sensors. Normally this should not present a problem.

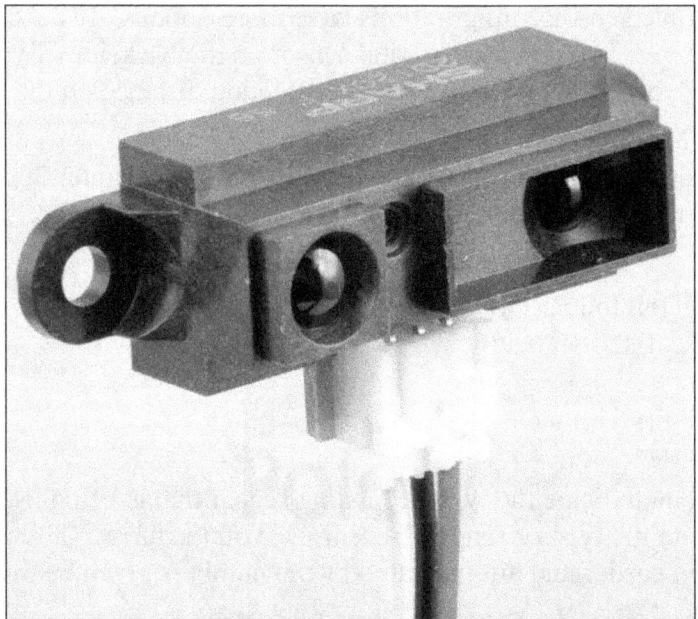

Figure 9.5: The Sharp IR sensors are easy to connect to the RROS chip.

The Ping Ranging Sensor

Figure 9.6 shows the Ping sensor. It too has 3 connecting terminals, two of which supply 5V and ground. The third terminal for each sensor will connect to the RROS chip as described later in this chapter. NOTE: The 5V supply for all of the perimeter/ranging sensors must be well regulated for best performance. If you have problems with erratic readings, try placing large capacitors (100uf, for example) from the 5V terminal to ground (observe polarity markings).

Figure 9.6: The Ping ranging sensor can be read with a digital pin.

The Ping sensors are read using a digital pin which, as we will see later, has its advantages. The drawback is that the timing loops needed to read the digital output take considerably more time than an analog reading. The RROS uses special techniques to read three alternate Ping sensors simultaneously so the longer time required should not be a problem for most applications.

The Maxbotics Ranging Sensors

The Maxbotics ranging sensors (see Figure 9.7) have the ability to indicate their measured distance serially, digitally, or with an analog voltage. The analog voltage method allows the fastest readings, so that is the only method the RROS will support.

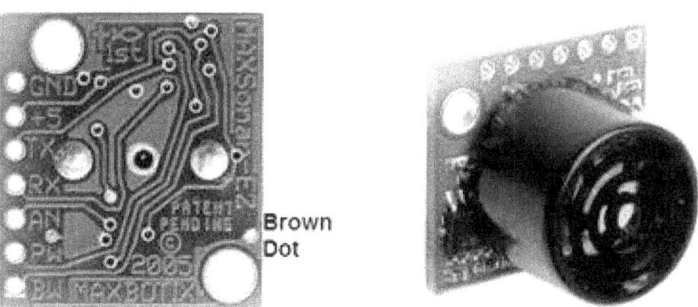

Figure 9.7: The Maxbotics ranging sensors have an
analog output mode that the RROS will support.

The left side of Figure 9.7 shows the connecting terminals for the Maxbotics sensors. The top two terminals will connect to ground and 5V. The **RX** terminal has to be pulsed to trigger the sensors. This synchronizes their operation so that sound waves generated by one unit will not interfere with the others. The **RX** terminal on *all* your Maxbotics sensors should connect to RROS Pin 12.

The terminal marked **AN** is the analog output terminal and should be connected to the RROS chip as described in the next section.

SR04 Sonic Rangers

The sonic rangers look very much like PING rangers, but have an incompatible, 4-pin interface. The power connections are 5V and ground. The output pin is labeled *echo*. A pin labeled *trig* on all five sensors must be connected together and then to a trigger pin on the RROS chip. These rangers offer a low-cost alternative to other ultrasonic ranging sensors. We import them and make them available at RobotBASIC.com.

Connecting the Perimeter Sensors to the RROS

Connecting the perimeter sensors to the RROS chips is more complicated than you might imagine. This is true because some of the sensor types require digital pins on the RROS chip while others require pins capable of AtoD conversions.

In the discussions that follow, we will refer to sensors and their related connections using the following abbreviations.

D	Digital Sensor
S	Short distance Sharp IR sensors (analog pin required)
L	Long distance Sharp IR sensor (analog pin required)
M	Maxbotics Ultrasonic sensor (analog pin required)
P	Ping ultrasonic sensor
R	SR04 Sonic Ranger
any	Any of the supported <u>analog</u> sensors may be used (but all sensors must be the same type)
tur	Turret connection (to CTL input on servomotor)
trig	Generates trigger pulse for Maxbotics sensors

Each abbreviation above will be followed by a number (if appropriate) to indicate the physical mounting position on the robot where that sensor should be placed. The numbers will correspond to the positions indicated in Figure 9.2. If the reference is to a turret mounted ranger, a **-ranger** will be used as the indicator.

You have many options for attaching sensors, because there are many sensor types with multiple configurations (turret, no turret, etc.). Let's look at each option in detail. Each begins with acceptable rCommand parameters.

Option 1: DIGITAL+IRSHORT
DIGITAL+IRLONG
DIGITAL+MAXBOTICS

In this mode, all the perimeter sensors are digital. There is no rear sensor. The range type refers to the type of sensor mounted on the turret. The connections to the RROS chips are as follows.

RROS PIN	SENSOR
15	D-1
16	D-2
17	D-3
18	D-4
12	D-5
14	any-ranger
10	tur

Option 2:   DIGITAL+PING

In this mode, all the perimeter sensors are digital. There is no rear sensor. The range sensor is a Ping. The connections to the RROS chips are as follows.

RROS PIN	SENSOR
15	D-1
16	D-2
17	D-3
18	D-4
14	D-5
12	P-ranger
10	tur

Option 3:   FIVERANGE+IRSHORT
                  FIVERANGE+IRLONG
                  FIVERANGE+MAXBOTICS

In this mode, there are five analog ranging sensors and one digital sensor on the rear of the robot. No turret is needed.

RROS PIN	SENSOR
15	any-1
16	any-2
17	any-3
18	any-4
14	any-5
12	trig (if Maxbotics are used)
10	D-6

Option 4:   FIVERANGE+PING

In this mode, there are five Ping sensors and one digital sensor on the rear of the robot. No turret is needed.

RROS PIN	SENSOR
15	P-1
16	P-2
17	P-3
18	P-4
12	P-5
10	D-6

Option 5:   SIXRANGE+IRSHORT
                  SIXRANGE+IRLONG
                  SIXRANGE+MAXBOTICS

In this mode, there are six analog ranging sensors. No turret is needed. No pin available for battery monitoring (discussed shortly.)

RROS PIN	SENSOR
15	any-1
16	any-2
17	any-3
18	any-4

14	any-5
25	any-6
12	trig (if Maxbotics are used)

Option 6:   SIXRANGE+PING
In this mode, there are six Ping sensors. No turret is needed.

RROS PIN	SENSOR
15	P-1
16	P-2
17	P-3
18	P-4
12	P-5
10	P-6

Option 7:   FIVERANGE+SONIC
In this mode, five of the low-cost sonic ranging sensors (available from our website) are used around the front of the robot to form the VSS. No turret is needed, but the rear bumper, if used, MUST be digital.

RROS PIN	SENSOR
15	R-1
16	R-2
17	R-3
18	R-4
12	R-5
10	trig
14	D-6

Reading the Battery Level
It may seem strange that discussing the hardware necessary to read the battery level is included in this chapter, but it has been included here for two reasons. First, the interface is so easy that it does not warrant a chapter of its own, but the second reason is probably the best reason for discussing it in this chapter.

Reading the battery level requires analog-to-digital (AtoD) conversion, as does all of the supported perimeter ranging sensors except the Pings and SR04s. The RROS chip has only six pins available with AtoD capability so there are not enough pins to monitor six analog sensors and the battery voltage.

One solution we provided was to support a rear digital sensor in some modes (thus eliminating the need for one AtoD pin). Since the robot should not normally back up any significant distance, a digital rear sensor should suffice for most situations. The primary drawback to using a digital sensor on the rear is that you cannot alter how far away it detects objects. Remember, if your robot needs a better view of what is behind it, one option is to just rotate 180° and carry out the necessary navigation using the full array of perimeter sensors on the front half of the robot.

If you want ranging sensors all the way around, one option is to just give up the ability to monitor the robot's battery voltage. If this is not a function you need, it is a very viable alternative. If you need battery monitoring and an analog rear sensor, we have provided a workaround. You can use the relay circuit described later to allow Pin 25 to read either the analog sensor for position 6, or the battery voltage.

If you think all of these options sound complicated, imagine how much trouble it was to weave all the possible options into the RROS code. The important thing is that the RROS can handle everything for you – no matter which option you choose. Just decide what sensors you wish to use, create an initialization subroutine to set them up properly and you are done.

Battery Monitoring

The battery monitoring function assumes that the battery voltage (or fraction thereof) will be read on RROS Pin 25. If you are using a digital perimeter sensor mode or if you are using a digital rear sensor or if you are using Ping sensors, then Pin 25 is available for battery monitoring and there are no conflicts.

If you are using six analog sensors though (IRshort, IRlong, or Maxbotics) the rear sensor is already using Pin 25 so you will need a relay to control which analog signal is actually routed to Pin 25. RROS will automatically control the relay when needed.

The *normal* position of the relay should select the rear sensor, so reading the perimeter sensors will be fast. When a request is made with the RobotBASIC command rChargeLevel() to determine the battery level, RROS will energize the relay, wait for it to switch, and then read the battery level. This should not be a problem, because one, the delay is minimal, and two, application programs should not need to read the battery level more than once per minute or so at the very most.

The relay circuit needed for battery monitoring will be discussed shortly.

Lowering the Battery Voltage

The RROS chip's AtoD conversion requires that the voltage being monitor be less than 5 volts – otherwise it could damage the chip. For that reason, we will reduce the main battery source to approximately 3-4 volts. If you have 6 volt battery you will need to reduce it to at least $^2/_3$ of its max. A 12 volt system will have to be reduced to $^1/_3$. This is easily accomplished with a couple of resistors or with a potentiometer. **NOTE:** If you prefer to use a potentiometer take care not to apply too high a voltage to the RROS input pin, as it can damage the chip. The voltage output from the dividing resistors will either connect directly to Pin 25 (if the rear sensor is digital or a Ping) or to a multiplexing relay that will route it to Pin 25, as shown in Figure 9.9.

Notice that the sensor output uses the Normally-Closed relay contact, so that the sensor is read when the relay is not energized. RROS Pin 10 will output a LOW (to energize the relay) when it needs to read the battery voltage.

If the Main Battery in Figure 9.8 is 12 volts, for example, **R1** could be 20K with **R2** being around 10K. If the Main Battery is 6 volts, **R1** could be 10K with **R2** being 10K to 20K. If your robot's battery is different, just choose resistors to produce a full-charge voltage between 3 and 4 volts. The actual number produced by the RROS chip is 0 to 255 with 255 representing a 5V reading. The simulator reports the battery voltage as a percentage of full – the RROS only provides a number that your programs have to interpret. **Note:** Early versions of the RROS reported ½ the actual value read, as we were trying to make it easier to get a value of 100 for a fully charged battery. Based on feedback from users, we have altered the RROS so that it now reports exactly what is read for the battery value.

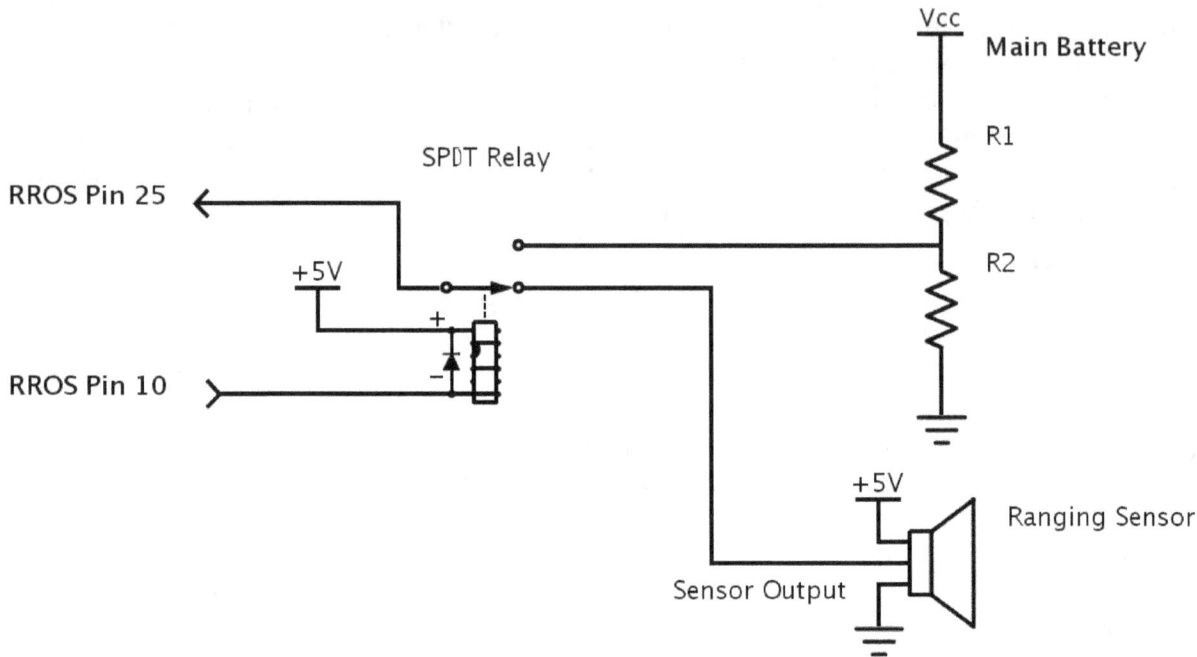

Figure 9.8: A relay is used to multiplex two signals to RROS Pin 25.

Reading the Battery

You can use the function rChargeLevel() to read the state of the battery. A fully charged battery should return a number based on your battery and the actual values used for R1 and R2. Although it is not really necessary, you could use a potentiometer for one of the resistors mentioned above, and adjust it to get an exact reading of 100 for a fully charged battery (regardless of the type or size of battery you are using), making the readings compatible with the simulator readings. As your battery depletes, the reading will decrease. You will have to experiment to determine what reading indicates that *your* battery should be charged. The number can vary considerably based on the type of battery you use and its age. How your robot reacts to this information is up to you (and your robot).

Calibrating the Turret Servomotor

When the RROS is initialized to use a turret, it will automatically attempt to center the servomotor and make all future movements relative to that center position. Since all servomotors are slightly different though, it is likely that your motor will be slightly askew. You can alter the center position with the command:

rCommand(SetTurretOffset, 128)

The number 128 in the command is the default position. Increasing and decreasing this number will change the servo's center position slightly. Once the turret is calibrated, using the command rRange(angle) will cause the RROS to move the turret and return the reading obtained from the ranging sensor. Remember, if you are using Virtual Sensors, the RROS will simply return the reading of the perimeter sensor closest to the requested angle.

Since most standard servomotors move approximately the same angular rotation for various control pulses, no other calibration should be necessary. In the rare case though, that your servo is non standard, or has a geared turret, we have provided the following command.

64

rCommand(SetTurretServoWidth,param)

The param in this statement controls how far the servomotor moves for each degree of turn requested in the rRange() statement. The default value is 16.

Calibrating the Virtual Sensors

If you are using one of the virtual sensor modes you can calibrate the ranging sensors to react to objects as you see fit. The following commands will set the distance (Param is in ½ inch increments) for the bumpers, read by rBumper(), and the proximity sensors, read by rFeel(). The *feel* distance on the simulated robot is approximately equal to its radius.

rCommand(SetBumpDist, Param)
rCommand(SetProxDist, Param)

Readings in Pixels

Normally, all range readings are provided in ½ inch increments. If you want the readings to maintain compatibility with the simulated robot, issue the following command:

rCommand(SetRobotDiameter, param)

where param is equal to the diameter of your real robot in ½ inch increments (enter 10 for a 5 inch robot). Any non zero param will make the RROS convert the normal ½ units to a pixel unit that makes the range readings the same as the simulator (relative to its diameter). For example, a distance reading of 80 means the object detected is twice as far away as the simulated robot's diameter. A reading of 80 on the real robot will mean the object is twice as far away as the REAL robot's diameter.

In Conclusion

This chapter conveys an enormous amount of information. Remember, it seems very complicated because there are so many options from which you can choose. Once you decide on the option that is right for your project though, the wiring necessary to connect the sensors described in this chapter to the RROS chip is actually pretty easy. Appendix B provides this information in a condensed form that may be easier for some readers.

You may be wondering why the pin-usage in this chapter is depicted in a variety of diagrams instead of just one simple layout. Unfortuanately, creating one simple layout is not possible. Let's look at one simple example involving the perimeter sensors to demonstrate this point.

If you use Maxbotic sensors they must be connected to pins capable of performing A/D conversion. Ping sensors on the other hand require digital pins. Since some pins on the RROS chip are digital only and others are analog only (still others can do both) the pin assignments for the perimeter sensors can vary considerably depending on the type of sensors you use.

Furthermore, since every pin is used on the RROS chip, whenever you change one pin assignment, several others may change also.

REMEMBER – all sensors have different characteristics. Ultrasonic sensors can be very accurate but they can only detect objects that reflect sound back toward the robot. This means that soft objects and objects with angular faces might be missed. The sound waves from one ultrasonic sensor might be reflected to another sensor (especially when the sensors are mounted close together, as they would be on a small robot). When this happens you get false and inaccurate readings. Ways to correct this are discussed in Appendix D.

IR sensors are nonlinear so their readings are generally less accurate. They cannot detect objects that do not reflect light back toward the robot (a black object, for example, might not be seen). Since IR beams are very thin, small objects that are between adjacent sensors can easily be missed. Ultrasonic sensors sometimes have an advantage here because of their more cone-shaped detection area.

A big part of building a robot is to choose sensors that are appropriate for your robot's environment and mounting the sensors to minimize interference and enhance detection. Many robot hobbyists have trouble with projects because of all of the above and fail to find solutions because they do not truly understand the problem.

Chapter 10

Line Sensors

Line sensors can be used in many ways. In our book, *Robot Programmer's Bonanza*, for example, we showed how the robot could find a short line on the floor that led to a battery charger. Following the line is an easy way to make sure the robot is in exactly the right place and oriented properly to connect to the charger. Properly calibrated line sensors can also be used to detect drop-offs so your robot won't tumble down stairs or off a table. You might even cover some of your sensors with a colored film so your robot could determine when it is over a specific color (maybe your kitchen has a distinctive colored floor, or perhaps you have taped a piece of red paper in from of the robot's charger. Always remember that nothing has to be used the way it was designed – use your imagination.

When line sensing is not needed, the rSense() function can be used to collect data for other uses. In Chapter 12, for example, we will see how it can be used to gather information that can help a robotic arm find and pickup objects. Our book, *Arlo: The Robot You've Always Wanted*, shows how to use the line-sensor inputs to detect the presence of people and animals.

Line Sensor Hardware

Obviously, the type of hardware you use for your line sensors depends on exactly how you are going to use them. When an actual line needs to be detected, you can use the QTR-1A from Pololu that was described in Chapter 7. The only requirement is that the floor and the line itself have enough contrast to trigger the sensor properly. For hobby situations, the "floor" could be a white poster board and the line created with black tape or magic marker.

In most cases, the distance between the sensor and the floor should be maintained as consistently as possible. In some cases it also helps to ensure that light from external sources cannot reach the sensor. This is not usually a problem as line sensors are often mounted under the robot's body (even though the simulator's sensors are slightly in front of it).

Line sensors MUST produce a digital output, which means they must produce a logical 1 or 0 (high or low voltage) depending on their detection state. In order to maintain true compatibility with the simulated robot, the line sensors should return a 1 when a line is detected.

Normally it will be assumed that you are using dark lines (none reflective) on a light (reflective) surface, which is perfect for the QTR-1A sensor because it provides a low voltage when close to a reflective surface.

Connecting Line Sensors to the RROS Chip

Normally, the RROS chip supports three line sensors (as does the default version of the simulator). Later chapters will explore methods of increasing the number of line sense inputs, but for now, there will only be three as shown in Figure 10.1

Figure 10.1: The standard RROS supports three line sensors as shown.

Sensor **L0** in Figure 10.1 is reported in the LSB of the data returned by rSense(), with **L1** and **L2** representing the next successive bits. The outputs from each sensor should connect to the RROS pins as shown in Figure 10.2.

RROS PIN	SENSOR
5	L0
6	L1
7	L2

Figure 10.2: The line sensors connect to the RROS pins shown.

Reversing the Logic

Depending on the type of sensor, and your application, you might prefer that the logic levels for the line sensors be reversed. This can be especially helpful when you have a white line on a dark surface, or the rSense() function is gathering data associated with some task other than line sensing. Because of this need, the RROS will allow you to set an *inversion mask* that can let you reverse the logic of any or all of the bits being read on the RROS's line sensor input pins.

Each bit in the inversion mask will correspond to the same bit position in the rSense() data. When a bit in the mask is a 1, the corresponding bit in the data will be inverted, allowing you to invert any or all of the bits using this command.

rCommand(SetInvMask, Param)

For example, you would set Param to 3 (binary 00000011) to invert the two least-significant bits of the data.

Chapter 11

The Compass

There are many novice-level projects that do *not* need a compass. Often though, the more advanced the behavior is, the more important a compass will be. When the robot can orient itself to face a particular wall in a room, for example, it has the potential of determining its location in the room by simply measuring the distance to each of the surrounding walls or objects.

A compass also provides the ability to measure the angles to beacons in the room, again providing the option of deter-mining the robot's position. More on this in Chapter 12.

The ability to navigate in a known environment can be greatly enhanced if your robot has a compass. Perhaps a simple example can demonstrate this point. Imagine the goal is to find your way to a specific object in a room while you are blindfolded. Before the blindfold is applied though, you get a chance to look around the room and mentally map where things are. During this period you could work out a path of sorts, where you use other objects in the room to give you feedback concerning your movements.

Once blindfolded you could turn an estimated amount and move toward the first object in your path list. When you reach it, you turn an appropriate amount and move to the next object in your mental map. Such a behavior would not be an easy task, but it is certainly within the realm of a hobby robot. It would make a great contest.

Suppose though, that after you are blindfolded, you are spun around several times. Without a compass, you would have no idea what direction you are facing, making it much harder to find your way to a specific object in the room.

Supported Compasses
Originally, the RROS only supported the Honeywell HMC6352 (previously available from Parallax, RobotShop, Sparkfun, and JameCo and others). The current RROS also supports the HMC5883. If you need an option not available on your chip, you may return your RROS chip to RobotBASIC and have it updated with the current version for a small handling charge. See the web page for details.

Interfacing the Compass
The RROS will communicate with the compass using the I^{2}C interface. The good news is that you do not need to know how the I^{2}C works. The RROS will handle all the details.

The compass chip itself is a small surface-mount IC that is difficult for hobbyists to use. Both the HMC6352 and the HMC5883 are offered in a variety of packages from various vendors. Refer to your documentation to determine which pins are which, on your package. A sample HMC6532 module manufactured by Parallax is shown in Figure 11.1.

Figure 11.1: This module is one of the ones we tested, but other compatible compass modules should work fine.

There are only four pins needed to interface with an I^2C compass. The V_{CC} or power pin, generally should connect to 5 volts (but always check the specs for your device). There should also be a ground pin that connects to the RROS ground. The final two pins are for the CLOCK and DATA. Connect the compass CLOCK pin to RROS pin 20 and the DATA pin to RROS pin 19. Both the CLOCK and DATA pins should have a 10K pull-up resistor from the pin to +5V (or whatever power supply your compass needs). IMPORTANT: The compass chip MUST be mount LEVEL in order to get accurate readings. The angular direction is not important though, as the RROS can compensate for that.

Notifying the RROS
You need to notify the RROS that a compass is present. This is done with the SensorSetup command. Just add the parameter HMC6352 or HMC5883 to the other sensor parameters. For example if your robot uses digital sensors and a PING ranger, you can indicate that plus the fact that it has a HMC5883 compass with the following command.

rCommand(SensorSetup,DIGITAL+PING+HMC5883)

You can also use an rCommand, as shown below, to set the compass type. This new option was added to make it easier to initialize special robots. Our **RB-9** and **Arlo** robots have become somewhat standardized, so special parameters in the rLocate statement can automatically initialize these robots so that you do not have to use many statements setting up speeds, sensors, motors, etc. This option does not, however, initialize a compass. If it did, and chose the wrong compass or any compass when none was present, then the RROS would hang in an endless loop. For that reason, a separate rCommand should be used to initialize the compass if one is available. This information is discussed in more detail in the Arlo book.

```
rLocate(ARLO,0)
rCommand(SetCompassType, HMC5883)
```

The above initializing methodology can be helpful even in you need to change a few parameter to those set automatically. Just use additional rCommands to alter the parameters in question.

Windows 8 Tablet Sensors
It is worth mentioning that you have other options when it comes to a compass. Most Windows 8 tablets have many sensors, including a compass, built in to their hardware. RobotBASIC provides

a utility for access these sensors. Refer to the book *Arlo: The Robot You've Always Wanted* (on Amazon.com, summer 2015) if you plan to use an on-board Windows 8 computer to control your robot. This is something we highly recommend, especially if you are building a powerful robot. Follow this link to watch a YouTube video showing Arlo in action.

http://youtu.be/ohpLRN-y2wY

Reading and Using the Compass

You can read the RROS compass angle using a statement like

```
dir = rCompass()
```

to find the current direction your robot is heading. You could use the compass to make your robot face East (90°) using the following code fragment.

```
while rCompass() <> 90
  rTurn 1
wend
```

If your robot is rotating at even a moderate speed though, it might overshoot the intended destination, so a better approach might be to have the loop end if the robot's angle is within a few degrees of the desired angle. The root of this problem is the fact that the compass can only be read every 80ms or so because of the Bluetooth delays (mentioned in Chapter 2). For these reasons we added a special command to the RROS to help you move the robot to a specific angle. This command will be discussed in more detail shortly.

Calibrating the Compass

Any compass will be affected by magnetic fields and metal objects in its vicinity so, for best results, you should calibrate your compass before using it in a new environment using the following code fragment.

```
SetTimeOut(40000)
rCommand(CalibrateCompass,0)
SetTimeOut(5000)
```

When the rCommand is executed, RROS will rotate the robot slowly for about 20 to 30 seconds while the compass is calibrated. Normally, RobotBASIC expects to hear back from the remote robot almost immediately (certainly within 5 seconds) so the first line above is necessary to prevent a Timeout Error. The last line resets the timeout period to the default period. You might offer a calibrate option in your applications through a button or menu item.

 Any electronic compass can be affected by magnetic fields (from audio equipment, for example) and even metal structures like a plant stand. For the best results calibrate your compass whenever it is used in a new location as well as anytime the accuracy seems to be a problem. You might also have to mount the compass above your main electronics using a short cable.

 The compass calibration is a good example of the power of the RROS chip. You don't have to worry about what compass you are using – just tell the RROS to calibrate it and it will perform the appropriate actions for whatever compass you've setup.

Setting the Robot Angle

Normally the compass should be mounted so that its magnetic directionality faces directly forward. It is certainly possible that your compass's mounting is a little off, or perhaps off by 90° or more because of how you had to mount a circuit board or breadboard holding your compass. If so, you could always add an appropriate constant in your application program, but we have provided a command that lets RROS handle that for you. Just supply a corrective angle (from 0 to 359) to be added to the actual reading. Due to the 8-bit limitation, you must divide the angle by two as shown below (and as we did with TurnToHalfAngle). This limits the correction precision, but only slightly.

rCommand(SetRoomAngle, angle/2)

The command above refers to the room-angle because this is probably an even more useful reason for offsetting the angle read by the compass. This simply means you can create an artificial North Pole. For example, you might want the compass to report Due North when it directly faces a particular wall in a room it is operating in. This makes it more like the simulation, which assumes that Due North means that the robot is directly facing the wall at the top of the screen.

Moving to a Specified Angle
The following command will move the robot to a specific angle, as discussed earlier. The angle in the statement can vary from 0-360°. Since the parameter is an 8-bit byte, the value must be divided by 2, thus allowing you to only specified even number degrees. This loss of accuracy is not generally a problem, especially since the movement is usually only accurate to $\pm 2°$ anyway.

rCommand(TurnToHalfAngle, angle/2)

The accuracy of this command is dependent on many factors, one of which is the accuracy of your compass as described in the next section.

Another factor is the speed that your robot moves toward the desired angle. Obviously you want it to move as fast as possible, but too much speed will make it difficult for the robot to actually stop on the desired location. For that reason, the following commands allow you to specify the maximum and minimum speeds (0-100) that the robot will use when it moves to its destination. Note: The actual speed will vary based on distance from the destination. The closer to the destination the slower the speed.

rCommand(SetTurnToMax, param)
rCommand(SetTurnToMin, param)

The default values for max and min are 60 and 15. The low speed should generally be the slowest speed that your robot will move without stalling. Experiment to find the right values for your situation. This is especially important because the speeds actually used will be slightly lower than the normal speeds because of the time needed to read the compass during the move.

IMPORTANT
There are many commands that utilize compass readings. The robot's ability to carry out these commands are based on the accuracy of the readings obtained from the compass. For maximum performance, it is important that your compass be mounted away from motors or anything that has magnetic fields or anything that can affect magnetic fields (metallic objects, for example).

Your robot's performance will also be affected by its environment. If there are objects in its environment that affect the compass readings (metal file cabinets, audio speakers, etc). For this reason, robust applications must utilize a variety of sensor readings to ensure the robot's perception of its environment is reasonably correct.

Chapter 12

The Beacon Detector

A beacon detector can be an extremely useful navigation tool. Our book *Robot Programmer's Bonanza,* for example, describes in detail how to navigate through a home or office environment using strategically placed beacons. *Robots in the Classroom* showed how a robot could triangulate on two beacons to find its position in a room, thus creating a Local Positioning System (LPS). We will summarize these concepts later in the chapter.

What is a Beacon

In general, a beacon can be anything a robot can locate and face. Our simulated robot uses colors to represent its beacons. Such an idea may sound strange, but if your robot has a camera you could actually use small disks of unusual colors for your beacons. The RROS, however, interfaces with a special chip capable of detecting infrared light pulsing at 56kh, which means all our beacons must be oscillating at that frequency. In order to create 15 different beacons, we will have each beacon periodically turn off its signal for a short time as shown in Figure 12.1.

Each beacon will generate a 56kh infrared signal for a repeating period of approximately 15ms. This time is indicated in the Figure as the ON TIME. This time period is not critical, but something close to 15ms is needed.

Each beacon will have an OFF TIME composed of a fixed initial offset plus a unique time associated with each Beacon. The RROS assumes the unique time for Beacon #1 is 500 microseconds, Beacon #2's is 1000 microseconds, Beacon #3's is 1500, etcetera.

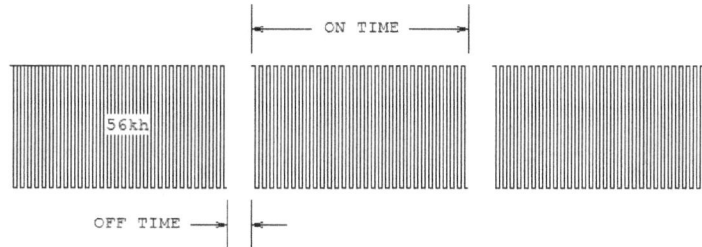

Figure 12.1: Each beacon generates this waveform.

Beacons can be built in many ways, the easiest of which involves a small micro controller. Controllers with reasonable speed can be programmed to generate the complete frequency patterns directly, including the 56kh signal itself. The timing for beacons is VERY sensitive so we offer a preprogrammed Beacon IC on our webpage that handles all the timing issues, making it easy to build your own beacons (See Figure 12.2). More on this later in the chapter.

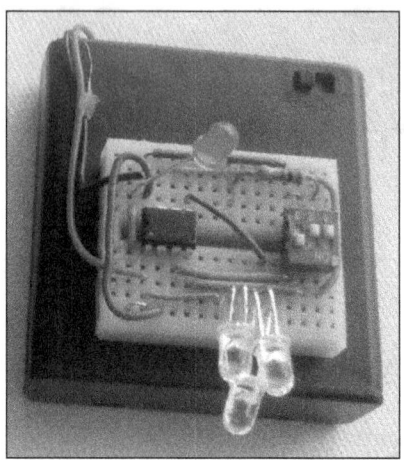

Figure 12.2: Our preprogrammed chip makes it easy to create your own beacons.

Each beacon will drive one or more IR diodes depending on your application. Multiple diodes may be required to ensure visibility from a variety of angles. You might want a beacon to be visible from nearly 360º, for example, if you hang a beacon in a doorway between two room, and want the robot to see it from any position in either room. Now that we know what a beacon is, we need a way to detect it.

The RROS Beacon Detector

The RROS assumes you are using the Vishay TSOP341 (Pololu #837) as the beacon detector, as shown in Figure 12.3.

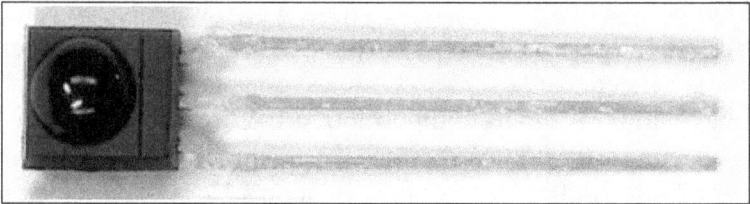

Figure 12.3: This small device serves as a beacon detector.

Connecting to the RROS Chip

We will connect the beacon detector to Pin 11 on the RROS chip through a 100 ohm resistor. This pin, if you recall from Chapter 2, is also used to drive the sound transducer. In Chapter 2 we connected the transducer in a quick and dirty way that made it easy to get started (the connection there did not require an inverter or other buffer), but that connection **MUST** be modified in order to connect both the sound transducer and the beacon detector to the same pin.

Rewire the transducer to the RROS chip and add the beacon detector as shown in Figure 12.4, then apply power to the chip and confirm that you still get the initialization tone when power is applied.

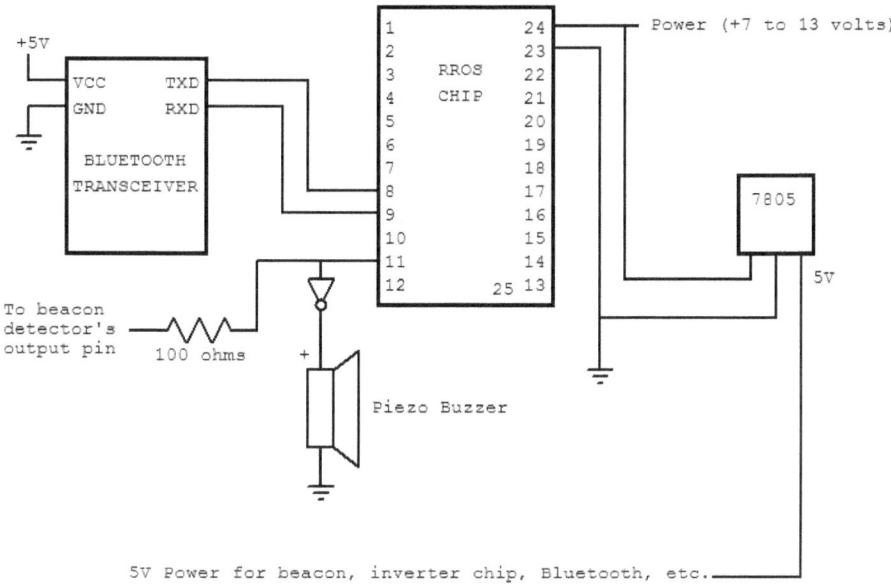

Figure 12.4: Rewire the sound transducer this way so the beacon detector can utilize the same I/O pin.

The beacon detector can now be connected as shown in Figure 12.5. Vishay recommends the resistor and capacitor shown to minimize noise problems.

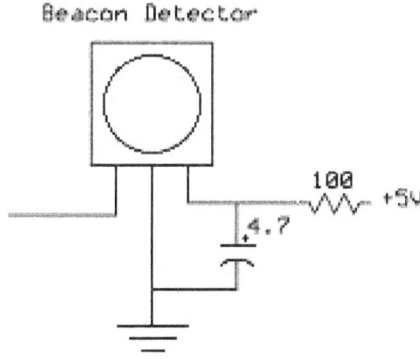

Figure 12.5: The output pin on the Beacon Detector(left-hand open pin above) connects to RROS Pin 12 through a second 100 ohm resistor.

The free pin on the detector (its output pin) should connect to RROS Pin 12 through a 100 ohm resistor. Make sure the transducer has be wired as shown in Figure 12.4. You can use nearly any inverter chip (such as one of the many variations of the 7404). Do not forget to apply 5V and ground to the inverter chip you use (refer to the spec sheet for your device for its pinouts). You can also use a transistor or other buffering device as we do on our RROS PCB (see Appendix D).

Once you have the hardware connected properly, the RROS software takes care of the details necessary to utilize both units on the same I/O line. In fact, as we will see shortly, having both of these items on the same pin will have an interesting advantage.

Narrowing the Detection Angle

For most applications, it is better if the detector only sees a beacon if it is directly in front of it. The reason for this will be discussed shortly. The easiest way to achieve this is to build a hood of sorts, that creates a small opening that prevents light from hitting the detector unless it enters directly from the front. Figure 12.6 shows the basic idea for the hood. Figure 12.7 shows one we built using black foam. It is important that you use something with minimal reflective properties as you do not want light to deflect down the channel walls giving you false readings.

Don't assume that all black materials will not reflect the IR light; in our experiments most did. The RROS tries to account for reflections when it can, but an appropriate hood will generally make things work better. The dimensions depend on your robot's environment (the power of your beacons, the distance from which they must be detected, etc), but an inch deep and a ¼ inch wide is a good place to start.

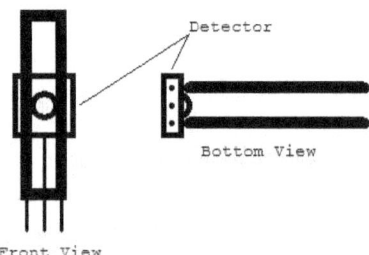

Figure 12.6: These two views help to visualize how to build a detector hood.

Figure 12.7: The hooded beacon detector can plug directly into a breadboard.

Detecting a Beacon

You can use the rBeacon(param) function to see if a beacon is detected. The param specifies the beacon number (or color number if using the simulator). The allowable range is 1-15 for the real robot, 0-15 for the simulation. For example, the following command

$$x = rBeacon(3)$$

will make x a 1 (or true) if beacon 3 is seen and 0 or false otherwise. **Note**: The simulator actually returns either zero or the distance to the beacon (which can be thought of as true). The RROS only supplies true/false information as it has no way of determining the distance to a beacon. A more robust (and more expensive) beacon sensor system could have been used, but for now at least, the distance feature is not available.

There is another difference with the RROS's version of rBeacon()compared to the simulator. If the parameter passed is zero (instead of a beacon number) then the RROS function will return true if *any* beacon is seen (the simulator would only look for black, which is color 0, in this case)). Any beacon, in this case, also means an IR LED flashing at 56kh regardless of its-off time.

As we will see shortly, it is a valuable robotic behavior for the robot to turn to face a beacon. This could be done using RobotBASIC commands, but it would be excruciatingly slow because (in order to be precise) the application program would have to read rBeacon() every time the robot rotated a degree or so and that generally means the robot can only turn at a very slow rate. To solve this dilemma, we added the following command:

$$rCommand(FindBeacon, param)$$

When the above command is executed, the RROS will rotate the robot until it finds a beacon. At that point the application program can use rBeacon() to determine *which* beacon was found. The value of param will determine which direction the robot rotates. A value of 0 indicates left while 1 means right. **Note**: rCommands always return 5 bytes in a string should you wish to use them. The last byte returned by the FindBeacon function is the number of the actual beacon found. Using this byte to determine which beacon was found is slightly faster than using rBeacon() since the data already resides within RobotBASIC.

There are two more rCommands that are useful during standard beacon operations.

Setting the Beacon Time

When you ask the robot to find a beacon, you do not want it to continue rotating forever if no beacon exists. Therefore, a time limit can be set that gives the robot a maximum time to look for a beacon. Generally, you should set the time limit to allow the robot to rotate at least 360° by changing the param in the following statement. The default value is 70.

$$rCommand(SetBeaconTime, param)$$

Muting the Beacon

Recall that both the sound transducer and the beacon detector are connected to the same RROS pin. This provides us the option of having the robot create a buzzing noise when the detector is pointing at a beacon – something that can be very helpful when troubleshooting beacon related behaviors. The default condition for this features is muted, but you can mute and unmute the sound by changing param to true or false (or 1 and 0) in the following statement.

$$rCommand(MuteBeacon, param)$$

Even when muted, the sound transducer will produce a small tick each time a beacon is detected.

Following a Beacon Path

Perhaps the easiest way to use beacons is to have them placed at strategic positions in the robot's environment. The idea is not complicated. The robot simply moves to fixed positions within the environment by finding and following the appropriate beacon signals. Let's start with a simple simulation situation as shown in Figure 12.8

Assume we know the robot is currently *somewhere* in the lower left corner of the room, and that we want it to move to Position 2 in the upper right corner of the room. Also assume that the rectangle in the center of the room represents chairs, tables, and other objects that the robot would have to avoid if it tried to take the shortest path to its destination.

A less complicated solution would be to have the robot move to Position 1 first, then move to Position 2, thus allowing it to take an uncluttered path. Once we know the destinations that we want, we can hang beacons on the walls or set them on cabinets or tables – any appropriate spot such that when the robot moves toward the beacon, it will pass over or near the desired destination.

Figure 12.9 shows how two Beacons can be positioned to allow the movements described above. The beacons should be mounted high enough in the room that the robot can see them over other objects of people. The upward viewing angle will change based on the distance from the beacon, which is why the vertical slit discussed earlier was recommended.

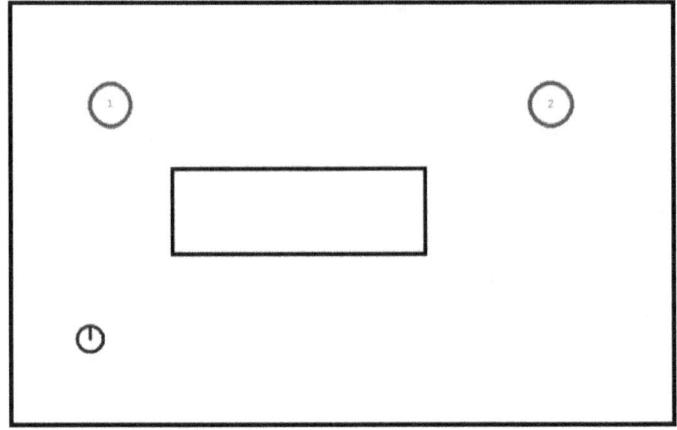

Figure 12.8: The robot can reach its desired destination
(Position 2) by passing through Position 1.

Notice the beacons are not necessarily *directly* behind the destination circles, because the robot, when looking for the beacons may see the outer edge and stop turning well before the center of the beacon is reached. This means that the turn-direction (CW or CCW) of the robot may be important.

The next step is to write a program that will make the robot look for Beacon 1, and move toward it until it gets a specified distance from the wall. At that point the robot should stop and look for Beacon 2. Once the beacon is found, it moves toward it until it again reaches an appropriate distance from the wall. As simple as this procedure is, imagine a series of properly placed beacons throughout your home or work place. A proper data table can allow your robot to move to any of the preset destinations by taking an appropriate path from beacon to beacon. This is a great way for your robot to find its way around your house, having one or more destination locations in each room. A detailed discussion of this idea can be found in our book *Robot Programmer's Bonanza* if you wish more information. For now though let's proceed with this simple example.

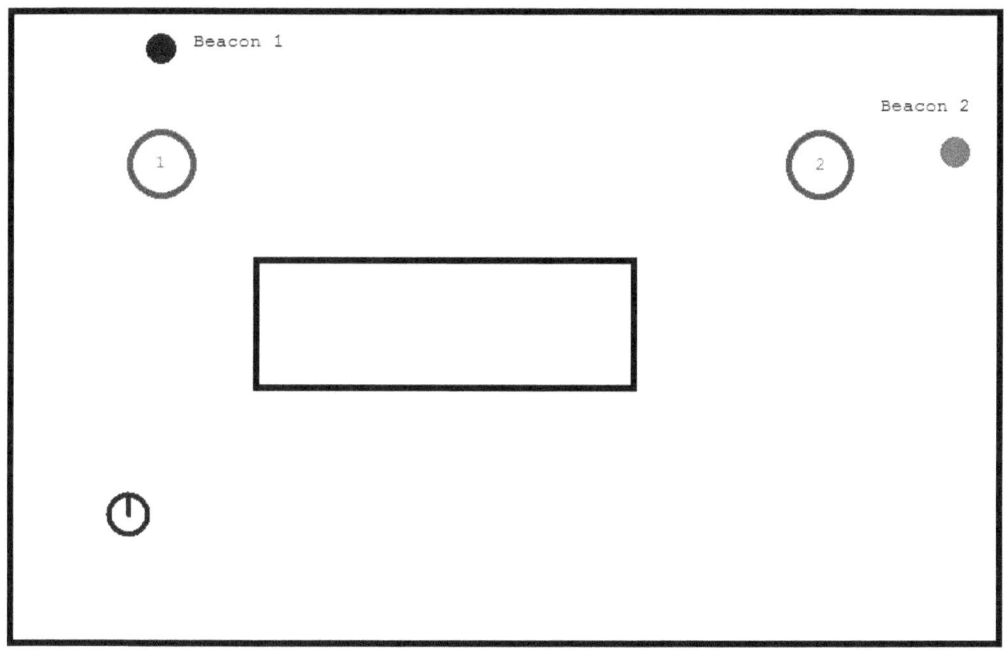

Figure 12.9: Properly placed beacons allow the robot to pass over
the destination circles when it moves towards the beacons.

The main Program in Figure 12.10 shows how the robot can be commanded to move to Position 1 and on to Position 2. Notice, that a special command (call) has been used in this program. Please read our HELP file whenever you are not familiar with a particular statement used throughout this manual. Some people like to own a hardcopy of the HELP file so they can read it on a lunch break or other situations where a computer is not necessarily available. For that reason we provide a 347 page book on Amazon.com, *The RobotBASIC Help File*, at a price cheaper than purchasing a new ink cartridge for your printer.

As you read the HELP file or manual, you will find RobotBASIC has hundreds of commands and functions not normally associated with standard BASIC dialects. The call statement is similar to the gosub statement used in most BASIC languages (including RobotBASIC), but call allows parameters to be passed. Also, the called modules, unlike standard subroutines, have local (instead of global) variables.

```
main:
  gosub Init
  call MoveToBeacon(1,100)
  call MoveToBeacon(2,120)
  rForward 0 // halt
end
```

Figure 12.10: The program demonstrates the logic necessary to reach Position 2.

Let's look at the logic of Figure 12.10. After initialization, the robot is simply asked to move to Beacon 1, then on to Beacon 2. The details of how it accomplishes these tasks are dedicated to the MoveToBeacon function. Notice that two parameters are passed to MoveToBeacon. The first of these, as you probably guessed, is the number of the beacon to find and follow. The second is the distance to the wall that controls when the robot should stop its movement towards the beacon.

The details of the MoveToBeacon module is shown in Figure 12.11. The first thing you should notice is how callable modules are defined. Instead of a simple label, they start with the sub statement and have a list of the variables being passed to them in parenthesis. The logic is straightforward. First, a second routine is called that forces the robot to face the desired beacon. That routine is also included in Figure 12.11.

Once the robot is facing the beacon, it moves forward turning toward the beacon when it does not see it and away from it when it does. This closed-loop feedback allows the robot to stay on course regardless of friction and wheel slip.

```
sub MoveToBeacon(BeaconNum,DistToStop)
 call FindBeacon(BeaconNum)
 while rRange()>DistToStop
  rForward 1
  if rBeacon(BeaconNum)
   rTurn 1
  else
   rTurn -1
  endif
 wend
return

Sub FindBeacon(b)
 while not rBeacon(b)
  // use the rCommand instead of rTurn for the real robot
  rTurn –1 // use for simulator
  // rCommand(FindBeacon,0)
 wend
return
```

Figure 12.11: Following the beacon is easier than you might imagine.

Of course, you still need the initialization module. One that works for the simulator is shown in Figure 12.12.

```
Init:
 // create beacons
 CircleWH 115,25,20,20,1,1
 xyString 150,20,"Beacon 1"
 CircleWH 750,105,20,20,2,2
 xyString 700,70,"Beacon 2"
 // destination circles
 Circle 100,100,150,150,Red
 Circle 625,100,675,150,Red
  // simulate clutter in room
 rectangle 200,200,500,300,Black
 // setup robot
 rLocate 100,400
 rInvisible 1,2,Red
return
```

Figure 12.12: This routine prepares the simulator's environment.

If you combine the modules and run the program, you will see the robot always ends up in the desired destination. Once you understand this principle you will find it to be an easy solution for controlling your robot's movements throughout a home or office environment.

Creating an LPS

The previous example is a viable solution for many navigation problems. Sometimes though you will need to know exactly where your robot is. We could use a GPS (Global Positioning System) like you use in your car, but unfortunately, the accuracy is typically several yards at best. What we need is an LPS (a Local Positioning System), capable of providing the coordinates of our robot within inches, not yards. In order to better explain the principles involved with building our own LPS, we will create one on the simulator.

For our simulation, we will use two beacons composed of specific colors. Before we start writing a program to use the beacons though we need to explore some mathematical principles associated with this project.

Figure 12.13 shows a simulated room containing the robot. Beacons are located in the two left corners of the room. We will see later how the robot can determine the angles to each beacon (as indicated in the Figure).

Think of the room as the 4th quadrant of a graph. The upper wall of the room is the X axis and the left wall is the Y axis. This makes the upper left corner the origin with the coordinates of (0,0).

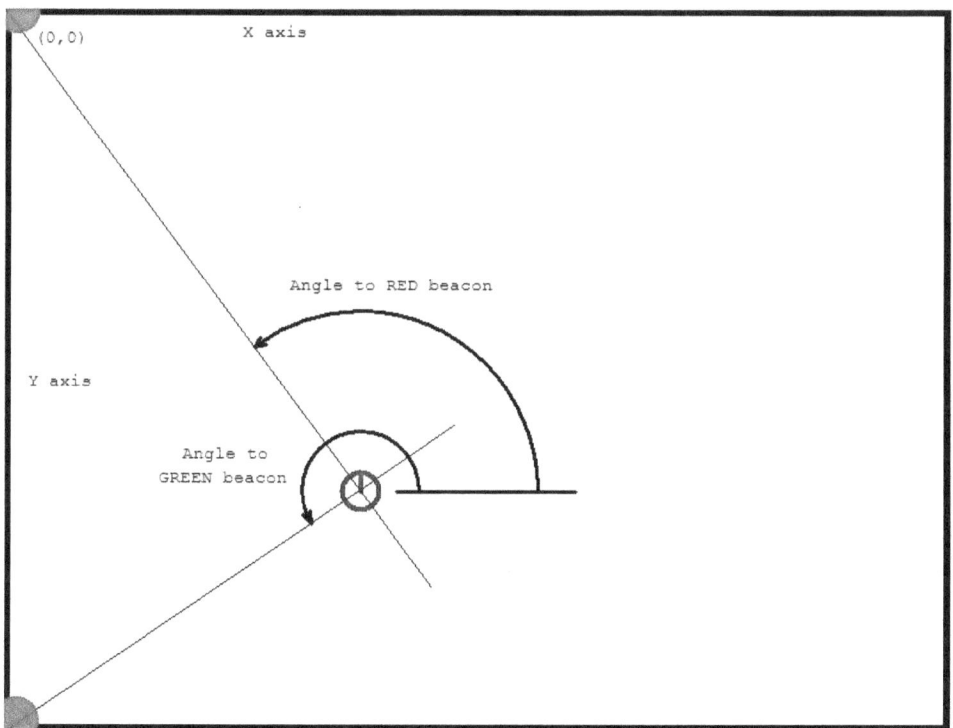

Figure 12.13: Two beacons can be seen in the left corners
of this simulated room (Red up, Green down).

Notice the two lines in Figure 12.13 that start at the beacons and intersect at the robot's location. If we could find the **x,y** coordinates of the intersection of these line, we would know where the robot is located.

Math is Essential

If you are going to study robotics, engineering, or programming, you should take as many math classes as you can, as mathematics is the language of technology. If you find the following discussion difficult, just skim it and concentrate on the final formulas. As with many mathematical situations, the final formulas derived from a principle are all that is needed to implement programs based on that principle.

The equation for any straight line will take on the following form:

$$y = m*x + b$$

In the above equation, **m** is the slope of the line and **b** is the point where it intersects with the y-axis. This means that we can write the equations for both of the lines in Figure 10.1 as shown below. The letters **r** and **g** (for red and green) will identify the parameters for the lines for each beacon.

$$yr = mr*xr+br$$
$$yg = mg*xr+bg$$

The slope of a line can be defined as the amount of change in **y**, for some change in **x**. This principle, is also defined in trigonometry as the tangent of the angle of the line, relative to a horizontal base, which is how the angles are defined in Figure 12.12 We will define the angle for the line to the green beacon as **Ag** and the angle to the red beacon as **Ar**. This means the two slopes can be defined as below.

$$mr = \tan(Ar)$$
$$mg = \tan(Ag)$$

The two lines intercept the Y-axis at the corners of the room. Normally, these intersections should be stated in feet or inches, but in our simulation we will use pixels. This means **br** = 0 and **bg** = -599, although, at some point, we will want to think of these numbers in terms of feet and inches. Our room is 800 by 600 pixels, so if we assumed it was a reasonable size, perhaps 20 feet by 15 feet, we can calculate that a foot is 40 pixels and an inch is 3.3 pixels.

If we substitute all of these new pixel values into the original equations for the lines, we get the following.

$$yr = \tan(Ar)*xr$$
$$yg = \tan(Ag)*xg-599$$

Remember, we want to find the point where these two lines intersect. At that point, **yr** will equal **yg** and **xr** will equal **xg**. Because of this, we will just refer to the intersection point as **x,y**. This means the equations can be rewritten for the intersection as follows.

$$y = \tan(Ar)*x$$
$$y = \tan(Ag)*x-599$$

Since the two equations are both equal to **y**, we can say that:

$$\tan(Ar)*x = \tan(Ag)*x-599$$

Solving the above equation for **x** gives:

$$x = 599/(\tan(Ag)-\tan(Ar)$$

Substituting the value of **x** back into the original equation for **yr** gives:

$$y = \tan(Ar)*x$$

When everything is calculated, the above equation will result in a negative number because the robot is in the 4th quadrant of the graph. If we are going to use the number calculated for **y** to locate our robot, though, we will have to multiply it by -1 because the distances down the computer screen are positive. This gives us two final equations for the location of our robot as shown below.

$$x = 599/(\tan(Ag)-\tan(Ar)$$
$$y = -\tan(Ar)*x$$

Using the Derived Equations

The equations derived above allow us to find the location of our robot if we know the angles from the robot to each of the beacons. Our next step is to develop a plan for finding these angles.

Our simulated robot (as well as a well-equipped RROS- controlled real robot) has a compass that can be accessed with the rCompass() function. The program in Figure 12.14 shows how to use this function and lets you see what values are provided as the robot rotates counter-clockwise.

```
rLocate 100,100
for i=1 to 10
 delay 2000
 print rCompass()
 rTurn -36
next
end
```

Figure 12.14: This program can help you understand the robot's compass.

If you run the program and watch carefully as it executes, you'll discover a few important things. First, the robot assumes straight up (perhaps thought of as due North) to be zero degrees. It also assumes that the angles increase as the robot turns clockwise, which is the opposite of standard graph notations (but appropriate for a compass heading), which were the assumptions made when we derived our formulas. Fortunately, we can transform these numbers with minimal effort using mathematics.

If we obtain the angle provided by the compass, we can convert it to a counter-clockwise rotation by simply subtracting it from 360. We can move the direction for zero degrees to the right by subtracting 270. This means the correct angle can be calculated as shown below.

Angle = 360-rCompass()-270

or just

Angle = 90-rCompass()

If the new angle is negative we can make it positive by adding 360° like this.

if Angle<0 then Angle=Angle+360

We can put all this into subroutines that allows the robot to calculate the angles, and then the robot's x,y position on the screen (or, in a room if it was a real robot). The subroutine FindAngles is shown in Figure 12.15.

```
FindAngles:
 // find them cclockwise first
 for a= 1 to 360
  if rBeacon(RED)
    Ar=90-rCompass()
    if Ar<0 then Ar=Ar+360
    break
  endif
  rTurn -1
 next
 for a= 1 to 360
  if rBeacon(GREEN)
    Ag=90-rCompass()
    if Ag<0 then Ag=Ag+360
    break
  endif
  rTurn -1
 next
 // save the angles
 TempAr=Ar
 TempAg=Ag
 //now find them clockwise
 for a= 1 to 360
  if rBeacon(RED)
    Ar=90-rCompass()
    if Ar<0 then Ar=Ar+360
    break
  endif
  rTurn 1
 next
 for a= 1 to 360
  if rBeacon(GREEN)
    Ag=90-rCompass()
    if Ag<0 then Ag=Ag+360
    break
  endif
  rTurn 1
 next
 // now average the angles
 Ar=(Ar+TempAr)/2
 Ag=(Ag+TempAg)/2
return
```

Figure 12.15: This subroutine determines the angles from the robot to the two beacons.

The code in Figure 12.15 is a bit longer than you might imagine, because each angle is actually found twice. Think about the beacons. They are not pinpoints of light so the robot will actually see the outer edge of the beacon before it turns directly towards the corner of the room (the center of the beacon), which is a source of error. If we find two angles to each beacon, approaching it from both clockwise and counter-clockwise directions, we can average those angles and get a more accurate answer.

Once Ar and Ag have been determined, the subroutine FindXY (see Figure 12.16) will utilize the formulas derived above to determine the probable position of the robot. FindXY also draws a vertical and horizontal line at the calculated position so you can see the accuracy of these routines.

```
FindXY:
 mg=tan(DtoR(Ag))
 mr=tan(DtoR(Ar))
 rx=599/(mg-mr)
 ry=-mr*rx
 line rx,0,rx,600,3,LightBlue
```

```
 line 0,ry,800,ry,3,LightBlue
return
```

Figure 12.16: This subroutine uses the angles to each beacon to find the location of the robot.

If you combine Figures 12.15 and 12.16 with the MainProgram and the Initialization module shown in Figure 12.17 you will have a program that moves the robot to 20 random locations and tests our equations by calculating where the program thinks the robot is. Furthermore, the program will compare the calculated positions to the actual positions and find the average error for the 20 tests.

Figure 12.18 shows the final output screen from the program. Notice that the average error was about six pixels, which would 2 inches in a 20 by 15 foot room. This error can increase considerably though, if angle measurements are faulty due to environmental conditions, but the error should still be under a foot even if the angles are off by several degrees. Experiment with your robot to determine your results.

```
MainProgram:
 xErr=0
 yErr=0
 for test=1 to 20
   gosub Initialization
   gosub FindAngles
   gosub FindXY
   xErr=xErr+abs(sx-rx)
   yErr=yErr+abs(sy-ry)
   delay 1000
 next
 xyString 200,10,"Ave x,y error = ",xErr/20,",",yErr/20
end

Initialization:
 LineWidth 6
 rectangle 2,2,797,597,Black,gray
 s=15
 c=Red\x=10\y=0
 circle x-s,y-s,x+s,y+s,c,c
 c=Green\y=590
 circle x-s,y-s,x+s,y+s,c,c
 sx=100+random(600)
 sy=100+random(450)
 rLocate sx,sy
return
```

Figure 12.17: These two modules complete
the program described in this chapter.

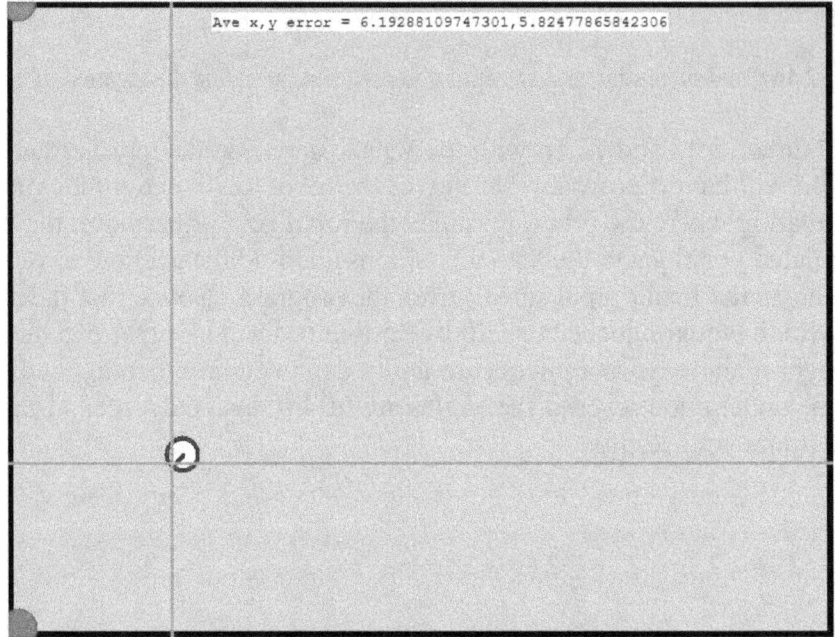

Ave x,y error = 6.19288109747301,5.82477865842306

Figure 12.18: The LPS simulation in this chapter produces this final output screen.

Potential Problems

It is important that the light from the two beacons does NOT overlap because the interference patterns generated will confuse the RROS detection software. Overlapping can happen when beacons are too close together, so you should consider having them at the ends of the longest wall in a rectangular room. It is also possible to have beacons that are too bright, although this is generally unlikely. If it happens though, the reflections can generate overlapping signals and cause misreads.

Making Your Own Beacons

If you wish to make your own beacons totally from scratch, you should consider using a small but fast processor programmed in C or Assembly. Creating a proper program is not trivial and may require significant amounts of effort. For example, when you create the 500 microseconds of delay for each beacon number you MUST also take into account the time delays created by the program code itself. For low beacon numbers this should not be a problem, but for higher numbers, the error accumulates causing your timing to be off.

Depending on your processor and your code design, you may also have an extra delay before the timing for the pause begins. In a perfect beacon program, the RROS should have to wait 250us (half the 500us beacon time) to start its count. This allows the readings to be taken as close to the middle of each period as possible, minimizing errors. If your processor takes time to start the pause period you can alter the delay the RROS uses with the following command.

rCommand(SetBeaconDelay, parameter)

The parameter (0-255) allows you to set a delay from 0 to 510us because the parameter you pass is automatically multiplied by two. If you are using a beacon you built, you will need to use this command to create an appropriate delay for your hardware. **Note:** The default RROS beacon delay is MUCH longer than this because of the way we built our beacon chips.

RobotBASIC Beacon Chips

The default pre-pause beacon delay is designed to work with a special beacon chip available on our web page. It is designed to make it easy to create your own beacons because you do not have to mess with all the critical timing. The chip is a small, preprogrammed microprocessor that comes in two varieties, one RED and one BLUE. More on this shortly.

Figure 12.19 shows the basic pin out for the beacon chip. You may use it to drive one or many IR LEDs. When driving only one or a few LEDs for experimental purposes, you can connect directly to the chip itself. If you need more LEDs and/or more distance you can use a transistor and drive multiple groups of LEDs as shown. The lower the current rating of your LEDs, the more you have to place in series to prevent burn-out, but the chip has been designed to work as shown with almost any IR LEDs. The use of transistors and higher current LEDs obviously will provide the greatest distances.

Figure 12.19 also shows an optional standard visible LED to serve as an ON/OFF indicator – very handy especially if you tend to forget to turn them off while experimenting.

Figure 12.19 shows +V for all the power connections. While you can use 5V, the chip allows you to power it with either 3 standard 1.5V batteries in series, or 4 rechargeable cells. Using batteries often makes sense if you are going to be moving the beacons around.

The number of LEDs you need will depend on your use of the beacon. If your robot will always be starting from nearly the same position (as you would in the example previously used) then you might get by with a single LED. If you want your robot to see the beacon from anywhere in a room, then you need to create a light source composed of several LEDs mounted at different angles. Also, the height of the beacon should be similar to the height of the beacon detector. For small test robots, this is easy. In other situations you will have to use your ingenuity. The beacon ship can drive a few LED's directly. If you need many though, a transistor buffer should b used (see Figure 12.19)

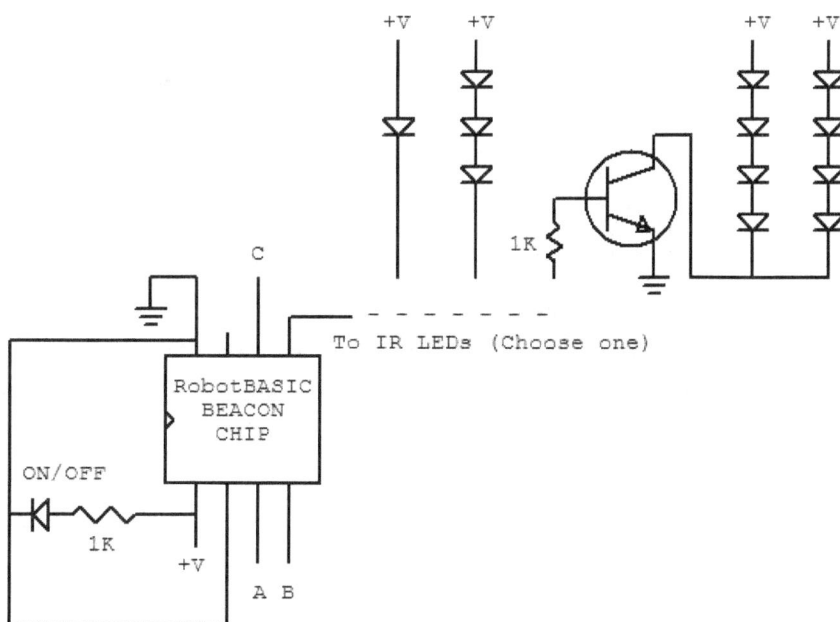

Figure 12.19: Our Beacon Chip makes it easy to build your own beacons.

87

If you are mounting your beacons at the top of your doorways, for example, and you have a man-sized robot, consider mounting the detector on a post as high as possible (see the tall robot in Figure 12.20). This not only puts the detector near the level of the beacons, but it also allows your robot to see over people's heads.

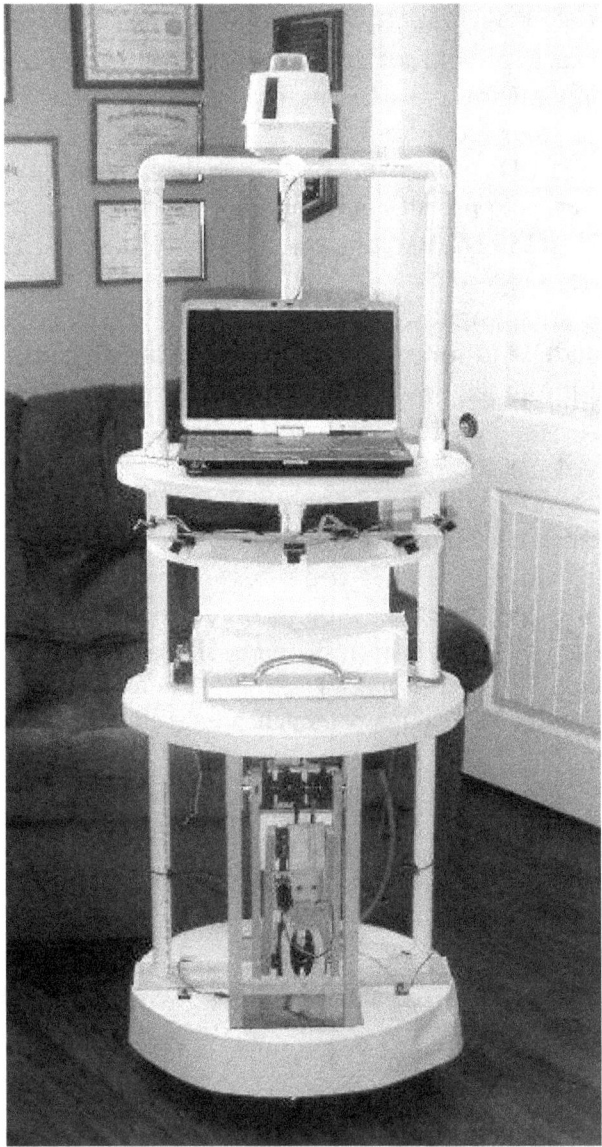

Figure 12.20: Note the high-mounted beacon detector.

Beacon Numbers

As mentioned earlier, the RobotBASIC Beacon Chips come in both RED and BLUE varieties. The pins labeled A,B and C in Figure 12.19 can be used to set a binary number indicating which beacon to create. Just ground a pin to make it 0 and leave it unconnected to make it a 1. These pins are constantly monitored and when changed, a new beacon number is implemented immediately.

The table in Figure 12.21 shows what beacon number will be created when pins ABC are connected as indicated. Red chips can become any beacon from 1 to 8 and a Blue chip can assume beacon identities from 8 to 15.

Most hobby projects will probably use only a few beacons so the limit of 15 different numbers should not be a problem. If your robot is operating in a large multi-room environment though, you many need many beacons, but you should be able utilize beacons with the same numbers as long as they are in different rooms. Of course, this means your robot will have to keep track of it current room location.

A B C	RED Beacon #	BLUE Beacon #
0 0 0	8	8
0 0 1	1	9
0 1 0	2	10
0 1 1	3	11
1 0 0	4	12
1 0 1	5	13
1 1 0	6	14
1 1 1	7	15

Figure 12.21: RobotBASIC beacon chips allow you to create 15 unique beacons.

Constructing Beacons

Our beacon chips make it easy to construct your own beacons as the examples in Figure 12.21 shows. The beacon on the right uses a solderless breadboard. The one on the left uses a solder-based prototyping board.

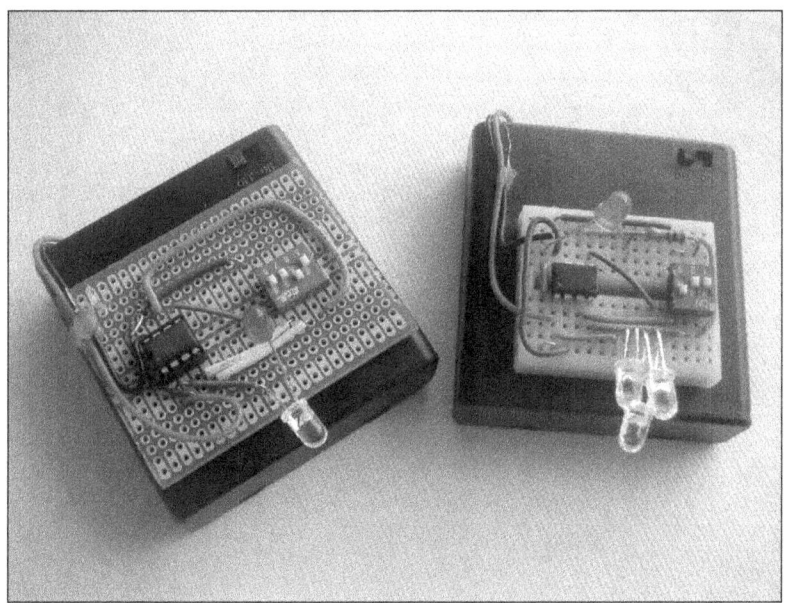

Figure 12.21: Two methods of constructing beacons.

The circuits for both beacons are mounted (using hot glue) on Radio Shack battery boxes that hold 4 AA cells. The boxes have integrated switches for turning them on and off.

The 4-gang boxes makes it easy to use 4 rechargable cells. If you wish to use standard AA batteries (of which you should only use 3 to get an acceptable voltage) just wrap an appropriately sized wooden dowel with tinfoil and insert it as the 4th battery.

In most cases you never need to change the number of the beacons you use (just tell the RROS what beacons you are using), so when you make a beacon you can just select its number by wiring the appropriate pins to ground. Since we were always experimenting, we used small DIP switches to make it easy to change beacon number for our beacons. Just connect each switch between a control pin and ground. When the switch is open the pin will act as a 1 and when the switch is closed (ON) the pin will be a 0.

You only need 3 switches for each beacon, but most DIP switches are generally sold in gangs of 4 or 8. Since we only had an 8-gang switch handy, we used a hacksaw to cut it into a 3 and a 4 gang, giving us the parts you see in the figure.

Chapter 13

The Arm Controller Expansion

The main RROS chip handles all of the standard interfacing needs of a mobile robot for both motors and sensors. We know, however, that advanced hobbyists will always want to add something we never thought of. Because of that we incorporated the ability for the RROS to interface with custom expansions. Chapter 14 will discuss the details of how you can create your own expansions. In this chapter we will look at an ARM controller expansion that we built into the RROS chip. The arm controller is a usable tool, but it also serves as an example of how custom expansions should work.

The RROS has a Dual Personality

In our original design for the RROS system, we envisioned it having the ability to control a robotic arm. Many factors influenced us to place the main RROS in a 24 pin chip which meant the arm subsystem would have to be dedicated to another chip in order to get enough I/O pins. The main chip had extra programming space, so we decided on a unique solution to our problem. We programmed the RROS chip to have a dual personality – it can act as *either* the main RROS or as a robot ARM controller.

Giving the RROS this ability provides extra value for the user. If two chips are purchased, they can either be used individually as RROS chips for two separate robots, or they can be used together with the second chip acting as an external arm expansion. Everything needed to make this happen is built into the RROS chip. In fact, our ARM expansion is just one of many ways you can expand the RROS. As mentioned earlier, details of how to create custom expansions will be covered in Chapter 14.

Connecting the Arm Expansion

Figure 13.1 shows how the ARM expansion chip should be connected to the main RROS chip. Every I/O pin on the main RROS chip is used, so we have to give up two rSense() inputs in order to gain two pins (for transmit and receive) to communicate with a second RROS chip acting as an Arm Controller. The functionality of the lost pins (plus more) will be provided by the expansion chip. The pins used for transmit and receive are labeled in the figure. Notice the required pull-up resistor (anything from 4.7-10K should work fine).

Of course, all the standard power connections apply to both chips and everything previously discussed for the RROS chip still applies.

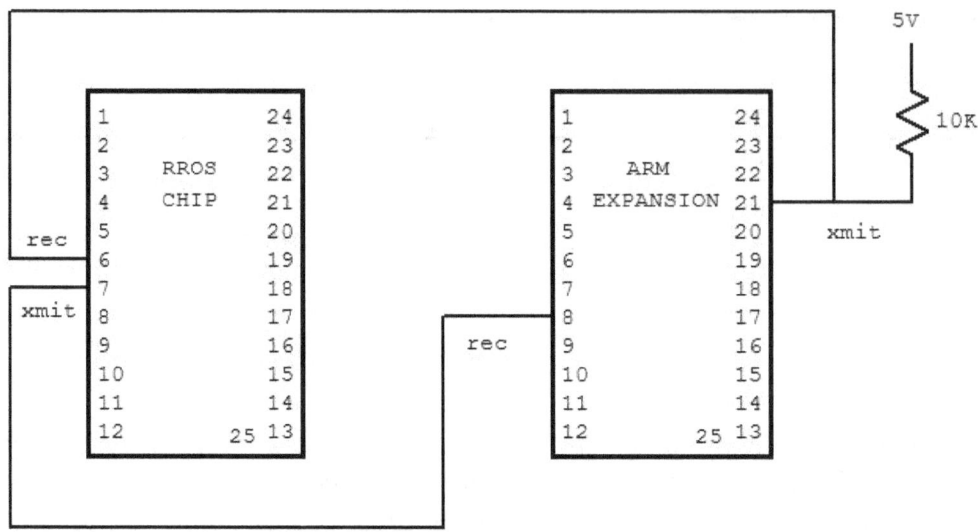

Figure 13.1: A second RROS chip can be connected as shown to act as an ARM Expansion.

The ARM expansion has many functions, but one of the primary ones is to drive five servomotors to power the joints of an arm. In addition, the expansion chip can gather *up to* six bits of digital data and five bytes of analog data to be used as sensory inputs associated with the arm and/or gripper (or any other needs your robot might have). It can even provide *up to* three digital outputs for controlling things like, relays, solenoids, etc. The ARM expansion has some additional functions available but they will be discussed later.

Initializing the Arm Expansion

Once a second RROS chip is wired to the main RROS chip we can set it up as the standard ARM expansion as follows:

        rCommand(ExpansionSetup, ARM)

The command tells the RROS that an arm expansion chip is available and to initialize it according to the parameter passed (in this case ARM – other options will be discussed later). There are other type of expansions: they will be discussed in Chapter 13. The actual processe that makes the remote RROS chip change personalities is very complex, but everything is automatically handled by the RROS.

Controlling Servomotors

The ARM controller can manipulate the speed and position of five servomotors to power the joints of a robot arm. Of course, you could use these servos for any application but their intended use is an arm. The following table lists where the control line for each servo should connect as well as the recommended function for each servo.

PIN #	FUNCTION	Servo #
3	shoulder	0
4	elbow	1
5	hand open/close	2
6	wrist up/down	3
7	wrist rotate	4

There are several advantages for using servomotors to power your arm. First, each servo can be independently instructed to move to a specific position within its range of movement, typically to a repeatable accuracy of a few degrees or less. There is *generally* no need to use sensors to determine where a joint is positioned, because the servomotor will move until the desired position is reached, unless of course the arm is blocked or over-loaded. If such situations are expected, you can add sensors to each joint (perhaps a potentiometer) and utilize the arm expansion's analog inputs to verify the arm's actual position.

The second advantage of using servomotors to power the arm joints is that servomotors can be purchased in a variety of sizes – everything from tiny micro servos to very large, high-torque monsters – and they all can be controlled with the same signals from the arm expansion chip.

Establishing a Servo's Parameters

You can specify a position parameter for each servo as well as a speed parameter that controls how fast it moves to the specified position. The following code fragment will set the desired position for each servo to the values in the array JointPos[]. Element 0 of the array is associated with servo 0, element 1 with servo 1, etc. The position parameter can vary from 0 to 255.

```
rCommand(SetServoIndex, 0)
for p = 0 to 4
 rCommand(SetServoPosition, JointPos[p])
next
```

The first rCommand establishes a servo index number. The second command uses the specified array element to establish a new desired position for the servo currently indexed. The RROS will *automatically* increment the index every time it is used. This causes the rCommand inside the loop to control servo 0, then servo 1, etcetera as the index increments. **Note**: After an operation on index 4, the index will move to back to 0.

You can of course, independently specify a new position for any specific joint, but in most cases, you will probably want to change the positions of all joints simultaneously. The ARM expansion chip will start moving each joint as soon as a new position is indicated, but in general, if you set all the joints as shown above, they will appear to all move in unison toward their destinations.

You can establish a speed for each of the servomotors with the array JointSpeed[] using a similar code fragment as shown below. The speed parameter can vary from 1 to 255 with 255 representing the fastest speed.

```
rCommand(SetServoIndex, 0)
for p = 0 to 4
 rCommand(SetServoSpeed, JointSpeed[p])
next
```

None of the servomotors are enabled when the expansion chip is first initialized. This was done so that you can provide the appropriate initial conditions for speed and position parameters before the motors become active. If the initial positional values you use are where you always *park* the arm before turning off your robot, then the arm will not have any spastic movements when it first becomes active. **Note:** Fast spastic movements occur when *any* servo controller is started with the positions of the motors not matching the starting positions assumed by the controller. As stated earlier, the solution is simple. Always park the arm in the same position before turning off the robot, then initialize the park positions before enabling the motors. The following command enables the motors on the RROS Arm Expansion.

rCommand(EnableServos, True)

Ideally the actual movement range of each servomotor should match the movement you need. Generally this will not be true, so you can alter the starting position for each servo by giving it a parameter as shown below (the current index is used to determine which servo is addressed just as in the previous examples). The parameter used can vary from 0 to 255 with the default being 50.

rCommand(SetServoMin, param)

You also have the option of adjusting the maximum width for each servo's pulse with the following command.

rCommand(SetServoMax, param)

The typical values for this parameter are from 0 to 100, allowing you to specify a percentage of the normal range. The default param is 50. Take care to not set either the MIN or MAX values beyond the physical limits of your arm or even the physical limits of the servos themselves. Since you can control both the minimum and maximum values for the pulse width to each servo, it is possible to utilize only a small portion of the servomotor's range, effectively increasing the resolution. When properly initialized, a position parameter of 0 should move a motor to its minimum allowed position and 255 should move it to the maximum allowed position. The arm controller will determine the actual pulses needed to move the arm within the range established with the rCommands.

As in previous chapters, the command parameters for the RROS rCommands are predefined variables. The actual values can be found in the include file or Appendix A if you prefer to use the numbers.

As mentioned, it is very important to establish and set the limiting parameters for your motors and arm to prevent potential damage. We suggest you do your initial experimenting with a test servomotor not attached to an arm or anything else in order to ensure you understand how all the parameters work. Then, connect ONE joint at a time to the expansion chip and establish its minimum and maximum values, as well as your desired PARK position. Generally you should then set the speed, min, max, and park position before enabling your motors. As stated earlier, if you always park your arm before terminating a program, initializing these parameters will eliminate the jerky startup motions servomotor-based arms have to make to move the arm from some unknown position to its startup default.

Reading the Analog Inputs

As mentioned earlier, the ARM expansion provides the ability to read five analog ports to be used any way you wish. The values are returned from an rCommand() function in a 5-byte string with the first byte representing analog port 0. The five analog signals you wish to monitor should be connected to the expansion chip pins as shown below.

ANALOG PORT	PIN #
0	18
1	19
2	20
3	13
4	25

The code fragment in Figure 13.2 demonstrates how to read the analog data and place that data into an integer array with the index of the array corresponding to the analog port number. The code fragment assumes that the array AnalogData[] already exists.

```
a = rCommand(ReadArmAnalogs, 0 )
for j=1 to 5
 AnalogData[j-1] = ascii(substring(a, j, 1))
next
```

Reading the Digital Inputs

In addition to the analog ports, the ARM expansion can have up to six digital bits connected to the pins shown below. The bit positions shown are automatically mapped into the corresponding bit positions of the rSense() data. Remember, position 0 (the LSB of the rSense data) is already occupied by the data on Pin 5 on the main RROS chip. This means you can have up to 7 bits of digital data available to be used for line sensors and/or for arm related sensors. **Note:** The rSense function, by default, can ONLY read three bits of line sensor information (because the simulator generally has only 3 line sensors). In this mode the rSense data received from the RROS chip will be limited to only 3 bits. The command rSenseType 5 tells the simulated robot to utilize 5 line sensors. More importantly for this situation is that this command also prevents the limiting of rSense data coming from the RROS. This command MUST be issued in order to receive all 7 bits as described above.

rSense() bit	PIN #
1	10
2	11
3	12
4	15
5	16
6	17

The combined bits (one from the main RROS chip and up to six from the expansion chip) are automatically returned when an rSense() command is used in the non-simulator mode.

Notice that the phrase *up to* six bits has been used above. The ARM expansion has been programmed to provide several options to give you the most flexibility possible. One of these options is to provide digital outputs so your programs can control any lights, solenoids, relays, etc. that you might wish to use on your robot.

Output Bits

The ARM expansion always has at least one output bit available on Pin 9 in the NORMAL arm mode just discussed. You can add two more output bits by giving up rSense() bits 5 and 6. You can instruct the RROS that this is your wish by initializing the ARM expansion chip with the following command instead of the one used earlier with the ARM parameter.

rCommand(ExpansionSetup, ARMwOUT)

You can establish values for the output pins by sending the proper parameter with the following command. A parameter of 6 (binary 110), for example, would set the upper two bits and clear the lower one. Any output bits not available in the current mode will be ignored.

rCommand(SetArmOutputs, param)

The Navigation Assist Mode

The ARM expansion chip has an additional mode that can be valuable for some applications. Chapter 11 discussed how the robot could use two beacons to triangulate its location in a room. To do so, the robot had to slowly rotate to find the angles to each beacon. The rotation had to be slow enough to ensure that the beacons would be spotted during the search. This periodic rotation can be exciting when performed by a small robot demonstrating the principles of triangulation.

A larger robot that has real-world tasks to solve, does not enhance its image, by having to stop and rotate every time it wants to determine its current location. A large robot also has problems moving in small increments because of its mass. In the long-run, perhaps a complete subsystem should be built that performs triangulation in the background. For now though, it makes sense to create an introductory version so that hobbyists can test it and improve upon it. For that reason, we added a Navigation Assist Mode to the ARM Expansion in order to help with these situations.

Nav-Assist Hardware Requirements

Our solution to the problem is relatively simple, at least in principle. Imagine a servo controlled rotating base (perhaps mounted above your robot's head) that contains two additional beacon detectors. One of the beacon detectors only detects light through a narrow slit as described in Chapter 11, Figure 11.6. The other detector uses a wider slit allowing the angle of detection to be greater. Let's see why using two detectors with these characteristics can be valuable.

If you only have one detector it must have a narrow field of vision in order to isolate the beacon angle properly. As mentioned earlier, this prevents a quick search for the beacon because a fast rotation speed can easily miss seeing a beacon during the search. The wide slit beacon detector allows the rotation speed to be much greater until the beacon is detected (by the wide angle detector). Once a beacon is seen, the speed can be reduced, letting the rotation continue until the beacon is spotted with the narrow slit detector. Working together in this manner allows both navigational beacons to be found in a *relatively* short period of time.

As you might expect, the software needed to manage this search process can be complicated, especially since the servomotor controlling the rotation must be precisely calibrated so that its angular position can be accurately calculated from the pulse width being used when a beacon is spotted. This is necessary because the relative angular position of the Nav-Assist beacons must be combined with the robot's compass reading to produce the true angles to each beacon.

All this sounds complex because it is, but that is the reason the RROS was designed to handle it for you. Your applications can simply request angular data for two beacons and the Nav-Assist system will find them and return the data to the application which will perform the calculations described in Chapter 11 to find the robot's location in the room. Before all this can happen properly though, we must establish a few parameters needed by the Nav-Assist system.

Initializing the Nav-Assist

First we must tell the ARM expansion chip that you want it to give up some of its normal functionality and add Nav-Assist capability. You do this with the following command.

rCommand(ExpansionSetup, ARMwNAV)

When the expansion chip is initialized in this way, it will use the 3 pins normally associated with the arm output bits to read the ouputs of the two beacons and control the Nav-Assist turret. This means that in this mode you do not have any output bits and you only have a total of 5 rSense()

input bits (four from the ARM expansion and one from the main RROS chip). You still have control over the five joint servomotors and the ability to read five analog ports.

Nav-Assist Pin Useage

The control signal for the servomotor for the Nav-Assist should connect to Pin 2 on the expansion chip. You MUST also run a 4.7K pull-up resistor between this pin and 5V. The output from the wide-slit beacon detector should connect to Pin 16 and the narrow-slit detector to Pin 17. All this may sound complicated, but the power provided by the ARM expansion chip is worth the effort needed to fully understand your options.

Constructing the Nav-Assist

The actual size and structure of your Nav-Assist turret can depend on the size of your robot. The photo's in Figure 13.2 through 13.5 show the important aspects of our version in hopes that it can stimulate your creative thoughts.

Our Nav-Assist assembly is shown sitting atop a small robot, almost looking like a head with a single eye. The dark slit in the front actually houses both the narrow and wide slit beacon detectors. Our construction projects nearly always use standard items you might around the home whenever possible. The main housing for our Nav-Assist assembly, for example, is a plastic butter container and the top is a coleslaw dish from a fast-food business. The top has no function of course, but it adds a little flair and would look even nicer with a few flashing lights inside. It might make a nice place to mount an electronic compass if the entire unit were to be made as a stand-alone assembly. The lower skirt is made from a rubbery shelf lining material. Since it turns with the assembly it provides a finished look.

Figure 13.3 shows the assembly with the rotating portion removed. In this picture it is easy to see how the skirt hangs down over the stationary wooden base.

The gear on the wooden base is hot-glued to a servo horn and is capable of about 180° of movement. Since it mates with the smaller gear (½ as many teeth) on the rotating assembly we get the required 360° of rotation.

Notice there is a hollow spindle in the center of the wooden base (made from the body of a ball-point pen). It was chosen because it was the perfect size to act as a bearing for the small gear to rotate on. The spindle is hollow, so the wiring for the beacon detections can pass through to the rotating section. It is important that these wires be VERY flexible to avoid binding as the upper assembly rotates. A good choice for wire is individual pieces of ribbon cable.

Figure 13.4 shows the upper assembly without the butter container cover. The slit assembly was made from balsa wood lined with black anti-static foam. Both beacon detectors are given a wide vertical view but the rear detector has a much narrower horizontal slot as depicted in Figure 13.5. Refer to Figure 13.4 to see how the front detector is mounted lower than the rear one, thus preventing it from blocking the rear detector's view (the dotted line shows the rear detector's view).

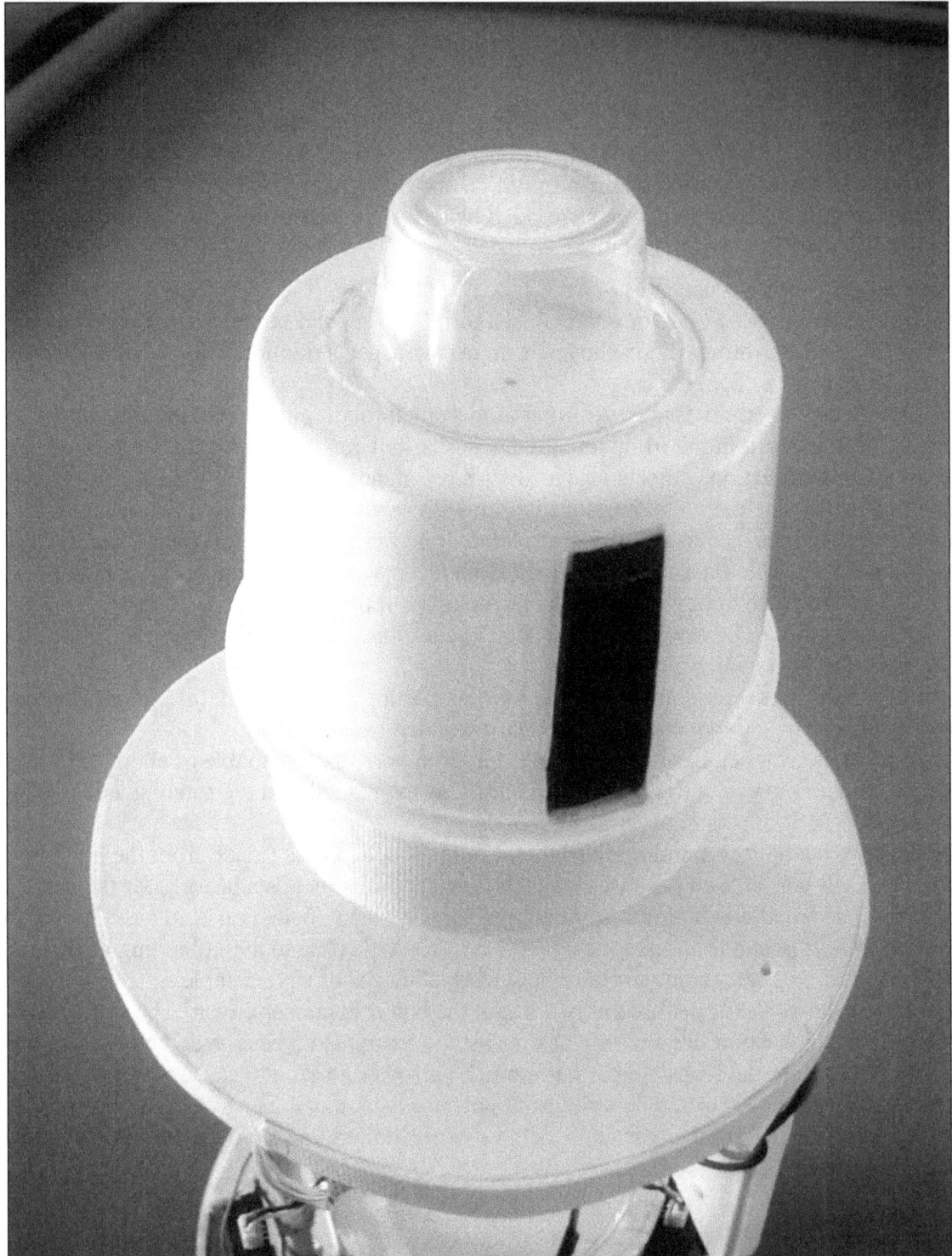

Figure 13.2: Our Nav-Assist assembly supports both narrow and wide slit beacon detectors that are rotated by a small servomotor.

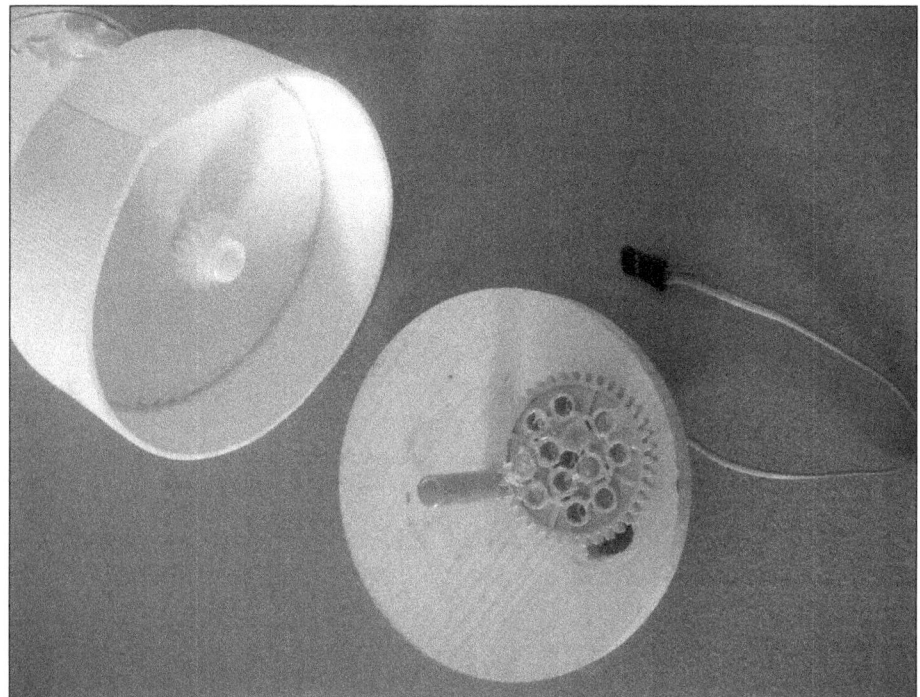

Figure 13.3: The upper assembly sits on a spindle and is rotated through two gears. The larger gear is attached to a servomotor.

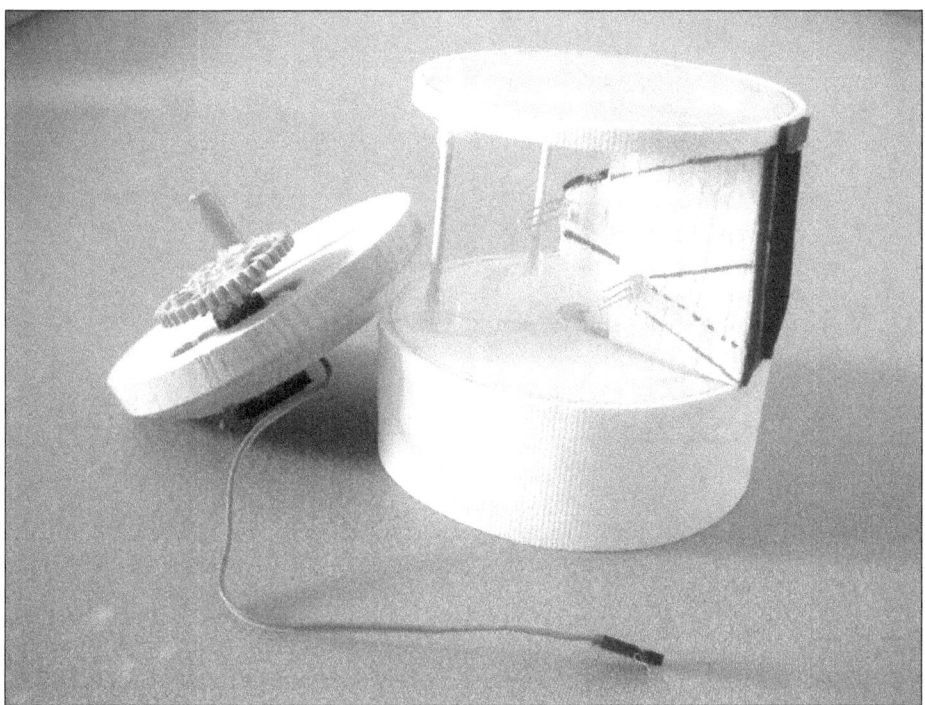

Figure 13.4: With the butter container removed, the slit-assembly that houses the detectors comes into view.

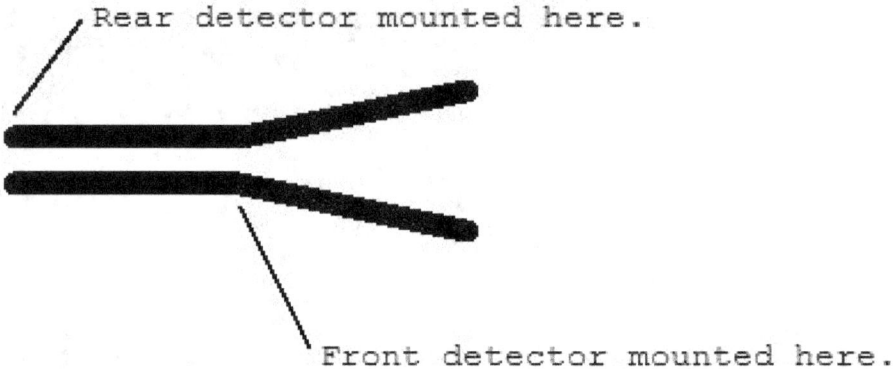

Figure 13.5: This top view of the slit assembly shows how the
two detectors can each have their own slit-controlled view.

Figure 13.6 shows the rotating assembly mounted on the wooden base with the servomotor clearly visible at the bottom. When the button container is added you get the finished looking product in Figure 13.2.

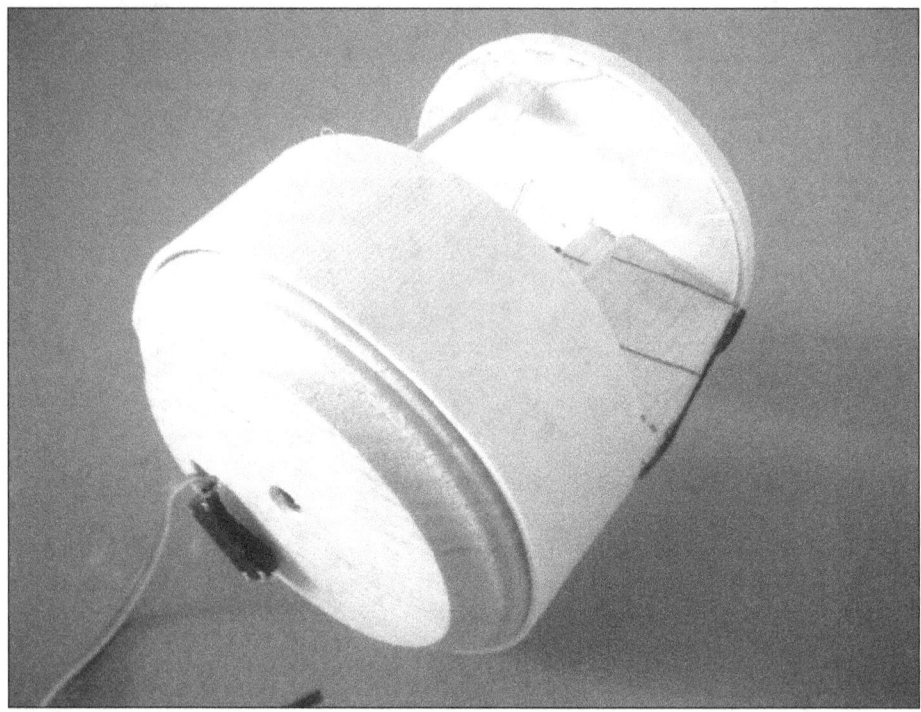

Figure 13.6: The skirt on the rotating assembly hangs down over the edge of the stationary base.

Consider It Beta

All of our construction, as well as the NAV-Assist software should be consider as a Beta Offering – that is it being offered to encourage testing and experimentation. In our basic tests, it works as described here, but we have not done testing in any real-world situations. As we receive feedback from users, we will pass those ideas along on our webpage and improve the software when we can.

Configuring the NAV-Assist

Once you have the beacon-seeking assembly built, it needs to be configured to operate properly. First, we need to tell it the beacon numbers for the two beacons that it should search for. You can do this with the following command. **Note**: the variable gr is the number of the "green" beacon and rd is the number of the "red" beacon as described in Chapter 11. This statement places the "green" number in the upper nibble and the "red" number in the lower nibble of the parameter being passed.

rCommand(SetNAVbeacons,(gr<<4) | (rd&15))

Calibrating the Servomotor

It is also imperative that the servomotor in the Nav-Assembly know the exact pulse widths needed at both ends of its 360° rotation. It will divide that range into 360 positions, so if the range is off, the reported angle will also be wrong. Since the pulse widths must be as accurate as possible we need more than 8-bit accuracy, so there are two commands used to set each value (one command for the lower byte and one for the upper byte).

In general, the lower range should be approximately 1000 and the upper range, 2000. You will have to experiment to find the numbers that will set an appropriate starting point for *your* Nav-turret, and move it exactly 360° (it should point at exactly the same position at each end of its travel). The commands below show how to set the two pulse widths (assume the lower pulse width is LPW and the upper is UPW).

```
rCommand(SetNavMinLow, LPW&255)
rCommand(SetNavMinHigh, LPW>>8)
rCommand(SetNavMaxLow, UPW&255)
rCommand(SetNavMaxHigh, UPW>>8)
```

Reading the NAV Angle Data

Once everything is configured and calibrated, the command ReadNAVangles will tell the Nav-Assist system to rotate the NAV turret, find the two specified beacons, and return the pulsewidth being used when the turret was pointed at each beacon. The first two bytes of the rCommand's returned data contains the green beacon information (high byte first) and the second two bytes contains the red beacon information.

Since we know the upper and lower limits of the turret's pulse width range (UPW, LPW) we can easily use this information and the angle reported from rCompass() to calculate the relative angles to both beacons. These calculations are easier if we assume the rCompass() angle has been calibrated with rCommand(SetRobotAngle, AngleCorrection) so that the room's "upper" wall is considered North. The following RobotBASIC commands demonstrate how to calculate the angles from the robot to the green beacon (grA) and to the red beacon (rdA).

```
a = rCompass()
r = rCommand(ReadNAVangles, 0)
gr=ascii(substring(r,1,1))<<8+ascii(substring(r,2,1))
rd=ascii(substring(r,3,1))<<8+ascii(substring(r,4,1))
grA = a+360*ga/(UPW-LPW)
if grA>360 then grA -= 360 //notice the C-style syntax
rdA = a+360*gr(UPW-LPW)
if rdA>360 then rdA -= 360
```

Once these beacon angles are obtained, they can be used with the equations in Chapter 11 to determine the robot's location in a room.

Potential Problems

As stated earlier, it is essential that the servomotor's end points are accurately defined. Another potential problem is beacon light reflecting down the channel. As mentioned in Chapter 11, this can be minimized if non-reflective materials are used to produce the assembly depicted in Figure 13.5. Unfortunately, just because a material is black does not mean it will not reflect the IR waves.

We know, from our tests, that some reflections are almost certain to occur so our RROS automatically records all readings and reports the average of those readings. The fifth byte returned from the rCommand() used retrieve the angle data will report the total number of readings recorded for both NAV beacons. This number should typically be less than ten in our tests. If it is much larger than this then your hood material is probably far more reflective than it should be.

We believe the NAV-Assist system has great potential. Please report the successes and failures you have, as well as your suggestions for improvements.

NAV-assist Programming Example

Figure 13.7 shows a SIMPLE program that acquires the parameters associated with beacon angles. Convert this data to angles as described in the text and use them to triangulate the robot's position in a room. Of course, you will have to use initialization routines and parameters appropriate for YOUR robot.

```
#Include "RROScommands.bas"
#Include "InitializationRoutines.bas"
gosub InitRROScommands
gosub InitRoboClawRobot
Main:
 rCommand(ExpansionSetup,ARMwNAV)
 rCommand(SetNAVbeacons,14+(3<<4))
 rSenseType 5
 low=600
 high=2100
 SetTimeOut 50000
 rCommand(SetNAVservoMinLow,low&255)
 rCommand(SetNAVservoMinHigh,low>>8);
 rCommand(SetNAVservoMaxLow,high&255)
 rCommand(SetNAVservoMaxHigh,high>>8)
 a=rCommand(ReadNAVangles,0)
 print (ascii(substring(a,1,1))<<8)+ascii(substring(a,2,1))
 print (ascii(substring(a,3,1))<<8)+ascii(substring(a,4,1))
 print ascii(substring(a,5))
end
```

Figure 13.7: This simple program demonstrates
how to access the NAV-assist data.

Chapter 14

Custom Expansions

In Chapter 13 we saw how an external ARM controller (made from a second RROS chip) could be added to the RROS system. In this chapter you will see how you can create your own expansions capable of adding new and enhanced features to the RROS. We have thought of a few types of expansion that many people might want and this chapter will discuss the RROS commands you can use to create those expansions. We also added commands to the RROS that allow you to add totally customized expansions so you can create new features that perhaps only you will need.

Expansion Communication

All expansions will communicate with the main RROS chip in exactly the same way, that is through a 9600 baud serial link as described in Chapter 13, Figure 13.1. It is important to realize that the expansion hardware can be anything that can communicate serially – a Parallax Basic Stamp, a Pololu Baby Orangutan, even a second copy of RobotBASIC running on the same machine as your application. This last option always surprises people but it is a viable option because RobotBASIC was designed so that several copies can be run on the same machine each having control over its own serial ports etc.

When the RROS chip wants to communicate with an expansion it will always send a 2-byte command and always expect a 5-byte answer. It is the responsibility of an expansion to watch for commands sent to it, and to respond appropriately when a command is received.

Types of Expansions

There are two basic types of expansions supported by the RROS – those that integrate their data into the standard simulator sensors and those that simply returns their data through rCommand() function. Let's begin by looking at an example of the second type.

Let's assume you have the need for an RFID sensor and you find that Parallax has one available on its web page. If you are not familiar with RFID sensors, visit the Parallax web page and read about them. You also see that Parallax provides code examples that show you how to use their sensor. This situation is exactly what we think our RROS should be able to handle and our goal is to allow you to cut and paste much of the code a company like Parallax might supply and only need to add a few extra lines of interfacing functionality in order to integrate their RFID sensor into your RROS controlled robot.

You could create any type of RFID system you need, but let's assume that you have altered the Parallax code so that your system can identify 20 different types of RFID tags that your robot will be using. Let's assume that you want to interface with your system with a command like this:

$$x = rCommand(ReadRFIDtag, 0)$$

In this command, the parameter is zero but it could be anything since a parameter is not needed for this application. We know that when an expansion receives the two bytes (the code for ReadRFIDtag and some parameter) that it must return five bytes to the RROS chip. Let's assume your code uses the Parallax supplied example code to make the RFID sensor check for the tags you are using. Let's assume you write code to set the value of the first byte returned to the RROS to zero (if none of the desired tags were detected) or to a number between 1 and 20 if a desired tag is found. That's it! That is all your system has to do. It simply monitors its serial input port for two byte sequences coming from the RROS, and when the first byte in the sequence is the correct command code (whatever code you assign to ReadRFIDtag) then it does whatever is necessary to read the RFID sensor and determine if an appropriate tag is present and send five bytes serially back to the RROS chip, of which the first byte contains the information you want. When the RROS gets these five bytes they will be echoed back to the current RobotBASIC application.

Your application program simply looks at the value of the first byte in the string x, and uses that information as necessary. You can use this basic technique to interface nearly any type of sensor, returning up to five bytes of data to your RobotBASIC applications.

There is one thing that is VERY important. You must choose an appropriate code to use for the command ReadRFIDtag. The RROS system has set aside codes 220 though 229 for use with custom expansions. Any rCommand that uses these codes will automatically have its 2byte sequence echoed over the serial expansion port. The RROS will then wait for five bytes to be returned, and echo those exact bytes back to RobotBASIC where it can be used by the RobotBASIC application.

Now that we have seen how custom expansions can return data through the string returned by an rCommand, let's examine a more seamless way of obtaining data from a custom expansion.

Unsupported Simulator Commands

If you have worked much with the RobotBASIC simulator, you know from reading the earlier chapters in this book, that there are several simulator commands that are not directly supported by the RROS chip. Since these are the most likely commands that users will want to implement, we have made provisions for anyone to create custom expansions to handle these situations. This feature is offered with the advanced hobbyist in mind, but we certainly encourage third party developers to design and distribute RROS expansions.

The RobotBASIC commands that are not directly supported are rPen, rLook, and rGPS. If the RROS detects one of these commands, and if a PEN, LOOK, or GPS expansion has been specified with an ExpansionSetup command, then the command will simply be echoed over the expansion serial bus. A custom expansion designed to handle one of these situations must simply watch for the 2-byte command sequence on the serial bus and, if seen, perform the desired operation and return five bytes containing the appropriate data. Let's discuss each of these situations independently.

A GPS Expansion

If you build a GPS expansion, you must tell the RROS that the expansion exists with the following command.

$$rCommand(ExpansionSetup, GPS)$$

Notice that this is the same method we used to tell the RROS we had the ARM Expansion. Next, you may want to send some information to the expansion itself – perhaps some initializing information such as what units to use for reporting coordinates. Your system may not need to do this at all – but the following command is available to you in case the GPS expansion *you* design needs any initializing parameter.

rCommand(GPSsetup, parameter)

Both bytes of this command will automatically be echoed over the serial link to *your* GPS controller. It is your controllers duty to watch for this command, and any others that it will need (more on this in a moment) and to respond to them appropriately.

The main command that a GPS expansion needs to handle is a request for data. In the case of the GPS, the command will be 66 followed by an unused parameter (typically 0). **Note:** Refer to Figure A-1, Appendix A for a list of simulator command codes. When this command is seen by the GPS expansion, it must do whatever is necessary to read the GPS it is using and return that data in an appropriate x,y coordinate form. For example, the units could be yards, meters, feet, etc. all relative to some pre-established position.

The x coordinate must be the first two bytes of the five returned bytes (MSB first) and the y coordinate must be in the next two bytes. These numbers will be reported when the command rGPS is used in an application and are always assumed to be positive. For that reason, your robot should *usually* be given a starting position other than 0,0. For example, if your robot's GPS uses feet as its unit of measurement and if you wanted it to move in an area 3000 ft by 1000 ft, then you could give it an initial coordinate reflecting where it currently resides. If it starts in the center of the space, for instance, then its initial position should be 1500,500. You can use the following reserved commands to set these initial values.

```
rCommand(SetGPShighX, 1500>>8)
rCommand(SetGPSlowX,1500&255)
rCommand(SetGPShighY,500>>8)
rCommand(SetGPSlowY,500&255)
```

Of course, your expansion must intercept these parameters and use them to establish the relative coordinates for your robot.

A PEN Expansion

Normally the simulator responds to the rPen command's parameter to determine if the pen should be raised or lowered. You can create a PEN expansion that watches for the rPen command code (129) and raises and lowers a real pen based on the value of the second parameter (perhaps performing this action with a solenoid or servomotor).

Even though there is no need for data to be returned, the expansion MUST send by five bytes (probably all zero) to adhere to the protocol. Remember, the RobotBASIC protocol always requires that commands are composed of two bytes with five bytes of returned data.

The RROS will handle all this for you as long as you tell it you are using a PEN expansion with this command.

rCommand(ExpansionSetup, PEN)

There is also a PenSetup command to perform any initialization, should you need it.

A LOOK Expansion
The simulator uses an rLook command (code 48 for angles to the left, and code 49 for angles to the right) to extract a color reading based on an angular direction specified. If your robot uses an embedded PC, perhaps the easiest way to create a LOOK expansion would be to run a second copy of RobotBASIC to act as the expansion, and have it communicate over a USB serial port with the main RROS chip as described in Figure 13.1, Chapter 13. Using RobotBASIC instead of some simple microcontroller has many advantages. For example, you could use RobotBASIC commands to take pictures from a Twain-compliant webcam (or from any almost any camera using RoboRealm, as described in our book *Hardware Interfacing with RobotBASIC*). Once the pictures are obtained, other RobotBASIC commands can be used to analyze the colors in the pictures making it easy to create a real-world version of the simulator's camera. Just return a color code from 0-255 in the *last* byte of the five bytes returned by the expansion as dictated by the RobotBASIC protocol (see the RobotBASIC HELP file for more detailed information). You can use the same color codes as the simulator or create some of your own. For example, you might have a special code for flesh tone colors.

The RROS will handle all the communication and reporting details as long as you tell it a LOOK Expansion is available with this command.

rCommand(ExpansionSeup, LOOK)

There is also a LookSetup command code reserved so you perform any initialization, should your LOOK system need it.

External Perimeter and Line Sensors
Even though we have tried to support the types of perimeter and line sensors that we feel a typical robot needs, there is always the possibility you want or need to do something different. For example, let's assume you have built a special camera system that can determine if there are objects close to your robot and even monitor the position of lines on the floor. After issuing the following command the RROS will assume that a special Sensor Expansion is available.

rCommand(ExpansionSetup, EXTSENSORS)

Once this command has been received, the RROS will constantly (nearly every time it receives a command) request data from your Sensor Expansion using the command code 202. Your External Sensor Expansion should watch for the code (as the first byte in a two byte sequence), and when is received, the expansion should create simulator compatible codes for rBumper, rFeel, and rSense and place them in the first, second, and third bytes of the five bytes it will return to the RROS chip using the serial communication link. These codes will automatically be ORed with the sensory data normally obtained by the RROS chip. This means that objects will be seen by your application if they are detected by *either* the expansion or by the RROS itself. Note: Any unused sensor data bytes should be set as zeros.

This is a very powerful concept that allows advanced sensory systems to integrate their data into the RobotBASIC standard format so it can be easily analyzed by any RROS controlled robot using the same commands as the RROS or the simulator.

There is also a ExtSensorSetup command to perform any initialization, should you need it.

Conclusion
The important concept in this chapter is that you can use the expansion principles discussed to expand and customize the RROS in a wide variety of ways.

As mentioned earlier, we certainly invite third party vendors to create expansions systems and make them available to RROS users.

Options and Limitations

Many people may only need one expansion, if they need any at all. We wanted to ensure though that advanced users have the option of utilizing multiple expansions in their projects. The following statement, for example, would tell the RROS that you have an ARM expansion, a PEN expansion and a LOOK expansion.

rCommand(ExpansionSetup, ARM+PEN+LOOK)

Having the option to have multiple expansions means there are some limitations that apply. For example, just as a particular expansion must be designed to recognize and respond to certain commands, it must also IGNORE all other commands. Expansions should not send back any data unless they are the intended receiver of the command initiating the transfer. This leads to another limitation.

Since there may be multiple expansions, we must insure that the serial communication has no conflicts. Since units only respond to their own commands, all units can be attached to the data line transmitting data *from* the RROS. The serial line sending data *to* the RROS however, is a different matter.

The 9600 baud transmission from an expansion to the RROS chip must send zero volts for a low, and FLOAT for a high, letting a single pull-up resistor ensure that the line is at 5V when appropriate. Notice the pull-up resistor back in Figure 13.1 used for this purpose. Most processors have the ability to float a line (for example, making an I/O PIN an input should effectively make that line a high-impedance load). If the serial port on your system cannot support floating the line for a high, then you must implement your own open-collector transistor driver to ensure that your expansion system does not pull the communication line low when it is not in use. We will not dwell on this point here, as we feel that most hobbyists wanting to create such advanced features will probably be familiar with this concept. Just remember that you have to handle this situation when two or more expansions are connected simultaneously.

Chapter 15

The RROS Robot Prototypes

During the design of the RROS, we needed to test each of the sensory configurations under a variety of situations and conditions. To help us accomplish that goal, we built numerous robot platforms compatible with our simulator, each with different motor and sensor configurations. In some cases the robots used one type of sensor during part of our tests and then later were rebuilt with different sensors. Our goal was to test as many configuration options as possible. We feel like these prototypes represent only a tiny sampling of what you can do with our RROS chips.

The purpose of this chapter is to discuss each of the robots in a way that might inspire you to create something similar, yet totally your own. Remember, the whole purpose of the RROS is that it makes it easy for nearly anyone to create a robot using almost any drive system and with a significant number of meaningful sensors, letting attention turn to programming intelligence instead of soldering and troubleshooting hardware. We have also striven to create a system that can be expanded and exploited by advanced users because we want to provide a system that can grow with you instead of becoming obsolete and outdated.

As you build your RROS-based robots, please let us know what features you enjoy and what features you think we missed. While we cannot guarantee to provide features desired by only a few, we certainly will try to provide reasonable capabilities wanted by a significant percentage of users. Creating RobotBASIC and the RROS chip has taken many years of effort. We hope you enjoy using it as much as we enjoyed creating it.

Construction Details

This chapter will concentrate on configuration details and software examples. Construction details and part numbers for items used to build the robots discussed in this chapter can be found throughout this manual, usually where the items are first mentioned.

It is important to realize that all of these robots could have been equipped with a full compliment of sensors as supported by the RROS. We certainly tried to test each sensor type with a variety of motor types and in various sensory configurations. In fact, we often tested various temporary sensory configurations on several of the robots discussed in this chapter before we settled on the robot's final design.

Using our RROS chip to build your robots means that you have many choices available to you. After you decide what you want your robot to do, you should determine its size, motor requirements, and sensory needs. No matter what you decide, we think in most cases that our RROS can make building your robot easier than any other way.

A Prebuilt Chassis

Many hobbyists want to build their own robot bases so what they have is unique or perhaps satisfies some specific requirements they have in mind. In many cases though, those that want to build a robot do not have access to the tools or even the space to handle the construction.

Fortunately there are many manufactures that offer prebuilt robot platforms in a wide variety of sizes. Unfortunately, most prebuilt platforms are very expensive. Figure 15.1 shows one we feel offers good value. It is the Dagu Rover 5 (the two motor model), available from Pololu. It provides a reasonable size, decent motors, and even wheel encoders for close to the cost of building everything from scratch. It comes with a battery holder for 6 AA cells, but more on that later.

Figure 15.1: This Rover 5 (Pololu #1551) provides a great pre-built base for your robot.

One of the nice things about the Rover 5 is that the wheel assemblies can be rotated to lower or raise the distance the body sits from the ground.

We built a balsa wood plate that mounts to the top of the Rover 5 as shown in Figure 15.2. Notice we added small wooden rails that add strength to the plate itself as well as creating a holding cell for the 6V, 4.5AH gel-cell battery. The motors are fairly powerful, but they are still small enough (barely) so they can be driven directly from the RROS chip. Most of our small prototypes were powered by six rechargeable AA batteries. The larger motors in the Rover 5 require more current, so a gel-cell battery such as the one shown in the figure is recommended. You might get by with standard AA cells, but rechargeable AA cells will NOT provide enough power for the motors and the RROS circuitry.

Notice that the balsa wood plate has room for two solderless breadboards providing the wiring space you need to begin experimenting with the RobotBASIC RROS. The wires protruding from the plate provide connections to the motors as well as the wheel encoders.

The new plate also provides appropriate places to mount perimeter sensors using either screws or hot-glue.

Figure 15.2: A custom top-plate holds the battery and two breadboards.

A Custom Small DC Motor Powered Robot

The chassis for the robot in Figure 15.3 was created from scratch primarily from foam board. It is powered with small DC motors driven directly from the RROS chip. Its sensors include short-range IR rangers configured as a VSS, an electronic compass, a beacon detector, battery monitor, and wheel encoders.

The robot in Figure xxx represents another inexpensive option. DC motors are often cheaper than servos, and the use of foam board means the only tool needed is an Xacto knife. Circuits can be built on a breadboard so you do not even need soldering skills. Notice the half-circle assembly in the figure that supports several of the IR sensors. The assembly rotates out of the way to make it easy to wire circuits on the breadboard.

Figure 15.3: This robot is powered by small DC motors connected directly
to the RROS chip. It has a full complement of sensors.

Sensors can be expensive, but you don't need to purchase them all at once. If you have to start with only one, you might choose the center IR ranging sensor. This would allow scanning of the area by rotating the robot as well as the ability to detect objects directly in front of the robot.

Another sensor to be considered for those on a budget is a beacon detector. They are very low cost and allow your robot to locate, and move to, a specific goal position in the room. The behavior for tracking to a beacon is also similar in many ways to following a line.

Remember, each robot you build should have its own initialization routine that establishes what motors are being used and what sensors are available. The routine should provide the parameters needed to ensure the robot moves in a straight line and turns the correct amount when asked as well as setting up any other parameters pertinent to it configuration. Below is a sample initialization file for this robot. Of course, the file could be longer or shorter based on your particular needs.

```
InitDCrobot:
 rCommmport 47 // change to YOUR port number
 delay 100   // needed by some Bluetooth devices
 SetTimeOut 10000
 rlocate 0,0
 gosub RROScommands
 rCommand(MotorSetup,SMALLDC+ENCODERS)
 rCommand(SensorSetup,IRSHORT+SIXRANGE+HMC6352)
 rCommand(SetMotorRamp,3)
 rCommand(SetClicksPerDiam,30)
 rCommand(SetClicksPer90,9)
 rCommand(SetReducForwRight,0)
 rCommand(SetReducForwLeft,5)
 rCommand(SetReducBackRight,0)
 rCommand(SetReducBackLeft,10)
 rCommand(SetRotationTime,2)
 rCommand(SetMoveTime,6)
 rCommand(SetSpeed,40)
 rCommand(SetSlowDownSpeed,22)
 rCommand(SetCCdivisor,3)
 rCommand(SetBumpDist,5)
 rCommand(SetProxDist,10
return
```

A Servomotor Powered Robot

Continuous-rotation servomotors include their own gear-boxes so they have plenty of torque and are often quieter than DC motors with external gearboxes. Our test robot, pictured in Figure 15.4 was built on a Parallax Boe-Bot base which was chosen because of its servomotor power train. A round foam board platform was added to provide space for circuits and sensors. The sensors on this robot include five digital IR sensors (shown below the outer edge), line sensors, wheel encoders, and a turret-mounted Ping))) ranging sensor.

The initialization file for a servomotor powered robot generally needs to contain all the things present in one for a DC motor robot, but it must also establish appropriate parameters specific to the servomotors themselves. For example, at-rest pulse width for the servos needs to be calibrated to avoid movement. It is also important to choose a maximum pulsewidth for the servomotors that produces linear control over the operational range. All of these topics are discussed in Chapter 4.

Figure 15.4: This prototype is powered by servomotors.

Below is a sample initialization file for the servomotor robot.

```
InitServoRobot:
 rCommport 44  // change to YOUR port number
 delay 100    // needed by some Bluetooth devices
 SetTimeOut 6000
 rlocate 0,0
 gosub RROScommands
 rCommand(MotorSetup,SERVOMOTORS+ENCODERS)
 rCommand(SensorSetup,DIGITAL+PING)
 rCommand(CalibServoDrive,0)
 rCommand(EnableCounters,1)
 rCommand(SetSpeed, 50)
 rCommand(SetSlowDownSpeed, 25)
 rCommand(SetRightStopOffset,135)
 rCommand(SetLeftStopOffset,126)
 rCommand(SetReducForwRight,5)
 rCommand(SetReducForwLeft,0)
 rCommand(SetReducBackRight,3)
 rCommand(SetReducBackLeft,0)
 rCommand(SetMotorRamp,10)
 rCommand(SetCCdivisor,30)
 rCommand(SetDriveServoDir,1)
 rCommand(SetDriveServoWidth,40)
return
```

A Steerable Robot

Chapter 7 explains how we modified a toy truck to act as the base for our steerable Robot prototype as shown in Figure 15.5. Even though the RobotBASIC simulated robot does not directly support conventional steering, we wanted to provide RROS support for this option for those that prefer or need it.

The entire purpose of this prototype was to test the steering and drive motor routines unique to this robot. For that reason, we did not permanently add any sensors to this prototype although it would certainly be easy to do.

Figure 15.5: Steerable platforms can be expensive, but we built this prototype from a toy truck.

Below is a sample initialization file for the steerable robot. With the steerable robot the left motor drive commands affect the rear-wheel drive motors and the right motor commands affect the steering servo. This means you can just set those parameters, but if you examine the RROScommands.bas file, you will see that these commands have retitled commands with identical codes, making their purpose more evident. For example, both SetRightStopOffset and SetSteerStopOffset are both equal to 117.

```
InitSteerableRobot:
 rCommport 44  // change to YOUR port number
 delay 100    // needed by some Bluetooth devices
 rlocate 0,0
 gosub RROScommands
 rCommand(MotorSetup,STEERABLE)
 rCommand(SensorSetup,DIGITAL)
 rCommand(SetSpeed, 60)
 rCommand(SetSlowDownSpeed, 35)
 rCommand(SetSteerStopOffset,135)
 rCommand(SetDriveStopOffset,126)
 rCommand(SetMotorRamp,10)
 rCommand(SetCCdivisor,20)
 rCommand(SetDriveServoDir,1)
 rCommand(SetDriveServoWidth,40)
return
```

The ability to control a steerable robot was added for those that require such a robot, even though it bears little resemblance to the simulator. It is important to emphasize that some commands either make no sense or simply are not compatible with this mode. For example, when the RROS calibrates the compass, the robot must rotate for about 30 seconds while the internal calibration takes place. The steerable robot does not have the ability to rotate so you must handle that situation yourself (perhaps you could rotate the robot manually while executing the command).

A Man-Sized Robot with Arm

We included a man-sized option as one of our prototypes because we wanted to demonstrate that our RROS could handle nearly anything a hobbyist might want to build. One of the interesting features on this robot is its arm which is placed in a very non conventional position as shown in Figure 15.6. One advantage of this configuration is that the arm folds away inside the body when not in use. Also, because of the placement of the shoulder joint, the gripper is able to obtain objects from a standard counter top or table and still reach items on the floor.

The need for large drive motors on our man-sized robot allowed us to develop the RROS code for interfacing with the RoboClaw motor controller as discussed in Chapter 5. This robot's primary sensors are Maxbotics Ultrasonic rangers organized as a VSS (see Chapter 9), wheel encoders, battery monitoring, an electronic compass, and a beacon detector. It also utilizes a second RROS chip that functions as an arm controller (see Chapter 13). The second RROS chip also controls a Navigation Assist System (also discussed in Chapter 13) for implementing a Local Positioning System as well as additional sensory inputs for the arm itself.

The Robot Arm

A RROS compatible arm, as discussed in Chapter 13, can have five servomotors for powering the shoulder, elbow, wrist up/down, wrist rotate, and hand open/close. Our robot arm provides all of these capabilities except for the wrist rotate function.

Commercial Arms

A search of the internet will provide you with numerous arms of various qualities and prices that are powered with servomotors, so they should work with our RROS. We had several design criteria that we wanted to utilize, so we decided to build our own arm.

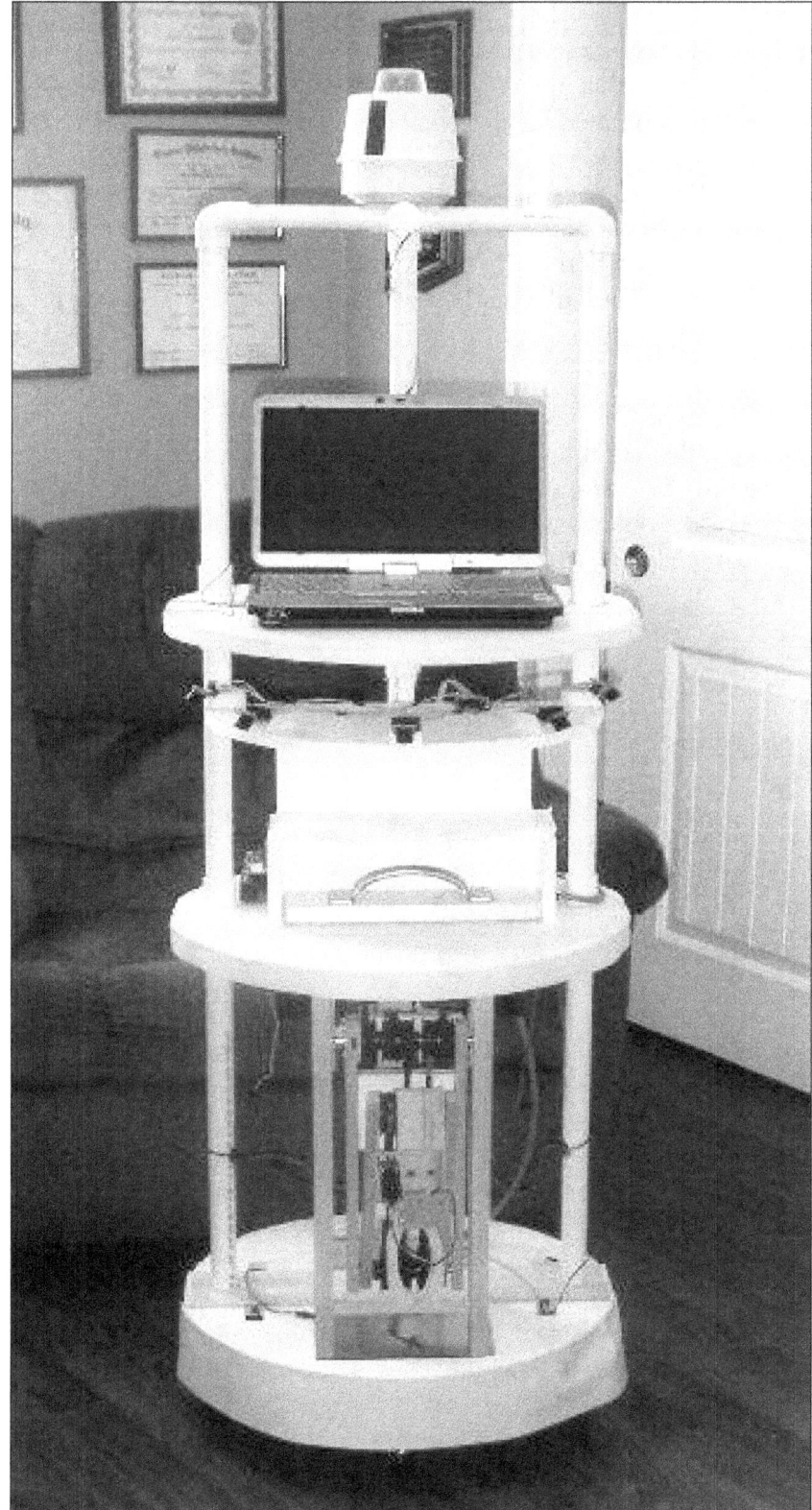

Figure 15.6: The RROS can handle robots of any size.

Our Arm

One of the unique features of our arm is that none of the joint-motors are mounted on the arm itself (although the small motor that opens and closes the gripper is mounted on the arm). It is rare to find this arrangement in a commercial arm even though there are some distinct advantages for it.

One advantage is that the arm does not have to lift the motors themselves, allowing the arm to handle much greater loads than would otherwise be possible. Perhaps the biggest advantage though, especially for a hobbyist, is that this arrangement makes controlling the arm far easier because once a joint is set to a particular angle, it will stay at that angle no matter how the other joints are manipulated.

The Belt Drive System

In order to mount the motors off the arm itself, we must have a system of belts and pulleys that transfers the motion from the motor to the desired joint. A rough approximation of our belt system is shown in Figure 15.7. Notice that there are three 2-gang pulleys, two on shaft 1 and one on shaft 2. The nature of the pulleys we used allowed us to attach two together with small bolts, but a high quality glue might work as well.

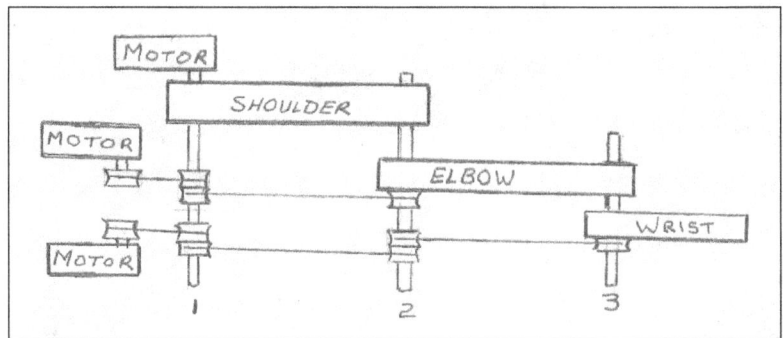

Figure 15.7: The motors powering the arm joints are mounted off the arm.

Let's examine the wrist joint as it is the most complicated. The wrist motor is connected by a belt to a 2-gang pulley which is connected to a second 2-gang with a second belt, which drives a final pulley (with a third belt) that is screwed to the wrist assembly.

All of the pulleys shown *on the shafts* MUST be the same diameter. This forces each joint to maintain its angular position no matter how the other joints are moved. You may use smaller pulleys on the motors if you wish, in order to increase the torque. If your arm is of any reasonable size and weight, you will probably need larger-than-normal servomotors – perhaps much larger depending on the dimensions of your arm and its intended payload. We kept our arm's weight to a minimum by using balsa wood and foam board when possible.

Our belts had teeth to prevent slippage that a v-belt or other alternative might allow. You could use chains and sprockets of course, but metal ones will add a lot of weight and plastic ones might not be strong enough.

Each joint needs some form of bearing to ensure smooth movements. Unfortunately standard bearings can add a lot of weight. We solved this problem by creating our own bearings from concentric brass tubes (available from most hobby stores) as shown in Figure 15.8.

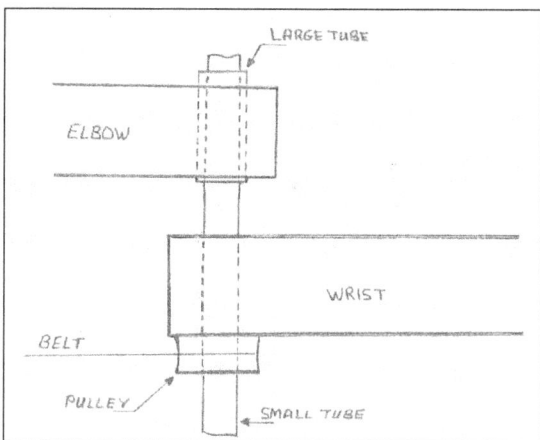

Figure 15.8: Concentric brass tubes serve as bearings for the arm joints.

The main axel for the wrist joint is a brass tube as shown in Figure 15.8. A slightly larger tube inserted into the elbow assembly serves as the bearing and provides minimal friction if it is lubricated with a small dab of Vaseline. If you cut your brass tubes with a pipe cutter, they will be slightly smaller at the cut due to the compression of the cutter. You will need to eliminate this on the larger tubes by filing the end of each tube from the inside with a round file.

The pulley in Figure 15.8 must be secured to the wrist assembly with glue or screws to ensure that the wrist assembly moves with the pulley. These rough drawings only show the basic principles of the arm's construction. An actual picture of the arm is shown in Figure 15.9. Notice the spring in the bottom left corner of the figure.

The spring in Figure 15.9 helps lift the shoulder joint as shown in Figure 15.10, which shows the back side of the arm where the motors are mounted on thin plywood plates. Notice the string wrapped around the pulley on the right side of the figure (our pulleys were wider than the belts allowing room to wrap the string without interfering with the belt – you could use a 3-gang pulley). That string extends to the spring mentioned earlier, allowing the spring to help rotate the pulley. The servomotor we used to power the shoulder joint was capable of lifting it without the spring, but the spring greatly reduced the strain on the motor. It also allowed the joint to remain in its last position without slipping even if the servomotors were turned off. Since the other joints are moving far less weight, none of them needed this type of assistance. We really debated about using a spring for assistance, but being able to turn off the servomotors after the arm is placed in a desired position, can often be a nice feature.

Figure 15.9: The arm folds into the body when not in use.

Figure 15.10: All joint motors are mounted behind the arm.

Figure 15.11 shows the elbow joint when viewed from the hand. Notice how the brass tubes create an axel between the shoulder assembly and the elbow assembly (shown closest to you). The single pulley on the right side of the figure is attached to the elbow assembly allowing them to move together.

The 2-gang pulley shown on the left side of the figure moves freely on its shaft using a second tube as a bearing. The right half of the pulley is driven by its motor through a second 2-gang pulley at the shoulder joint. The left side of the ganged pulley in the figure drives the wrist pulley as shown in Figure 15.12.

Figure 15.11: Notice the 2-gang pulley (on the left) in this elbow joint.

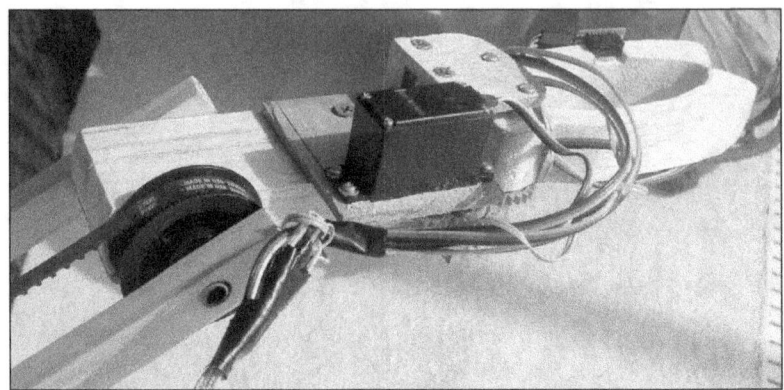

Figure 15.12: The hand uses a standard servomotor to open and close the balsa wood fingers.

Figure 15.13 shows the hand itself. Notice the small gears at the rotating base of each finger. Since they are connected, when one finger moves, the other also moves, but in the opposite direction, so they open, or close, together. One of these gears is driven from a gear on the servomotor's output shaft.

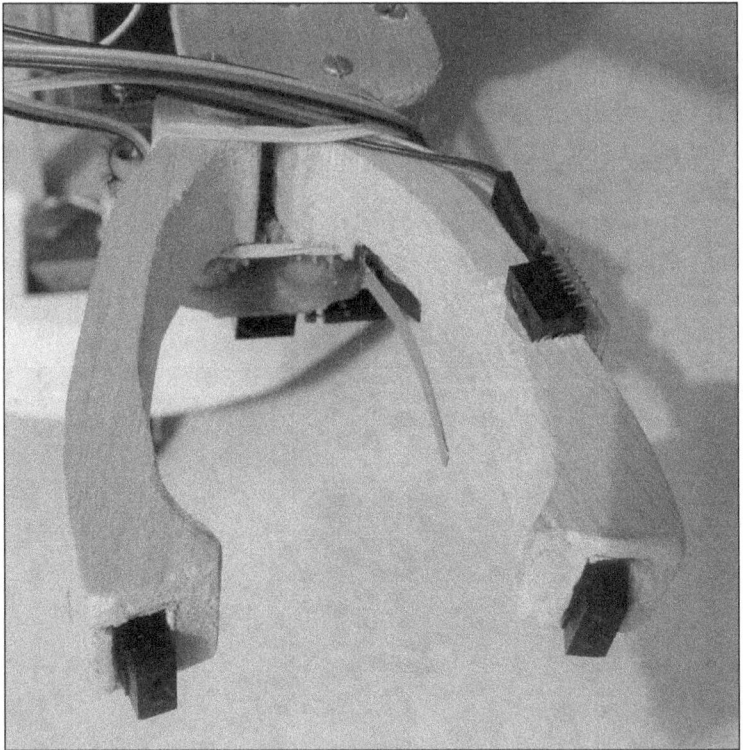

Figure 15.13: The fingers have several sensors.

Notice the sensors on the fingers. Two digital IR proximity sensors in the ends of each finger can help your programs to determine when the object to be picked up is centered between the fingers (neither sensor is triggered). You could then move the arm (or the robot itself) forward until the third IR sensor (shown on the finger in the right side of the figure) senses that something is actually in the hand between the fingers.

Once your program knows an object is within the confines of the hand, the hand can then be closed until the long lever of a snap-action switch is pressed (then close it a little more to get a good grip). The lever of this switch can be seen in the figure. There is also a digital IR sensor located under back of the fingers pointing downward to enable programs to determine when the hand is near a surface (such as a table). Notice also the rubber band around the two fingers (in the back). This helps provide a stronger grip much like the spring helped lift the shoulder joint.

Using the Sensors

The hand sensors can be connected to the sensory inputs on the Arm Expansion chip (see Chapter 13) allowing them to be read with the rSense() function. In order to use the arm intelligently, you must write programs that utilize its senses to dictate its actions. Let's examine this idea briefly.

In addition to the hand sensors discussed already, let's assume that your arm also has a small webcam mounted on the wrist so it can see objects immediately in front of the fingers. **Note:** We describe options for obtaining color information from a camera in our book *Hardware Interfacing with RobotBASIC*. Of course, this means that the PC to which the webcam is attached must be mounted on the robot (see Figure 15.6). There are other advantages to having an integrated PC. It can allow your robot to speak and understand voice commands (also discussed in the book above). You could even consider using RobotBASIC's graphics to create a face on the laptop's screen to give your robot additional personality.

Let's create a simple goal for the robot to demonstrate how it might be accomplished. Suppose there are three small objects (each of a different color) sitting on the floor in front of your robot and you order the robot to pick up the RED object.

Assuming the arm is currently folded away in the body, it could move the wrist joint alone to point the camera at the general area in front of the robot and acquire images. The images could be analyzed with RobotBASIC's *vision* commands to find the x,y position of the desired color within the image. Using the x coordinate, the robot could rotate left or right until the color is seen in the center of the screen (horizontally). It could use the y coordinate to determine how far away the object is.

The robot could move itself some predetermined position from the object in question, then move the arm so that the hand is horizontal to the floor, positioned only a few inches above it (this action could use predetermined positions for each joint). The arm could then be lowered until the bottom sensor on the hand senses the floor – confirming that the arm is positioned at an appropriate height. The desired color should still be in the center of any captured images, but the robot could rotate left or right to ensure this.

With the fingers closed, the robot could move forward towards the color (moving left and right if necessary to keep the color centered in the webcam's view). Since the fingers are the same height as the objects, and since the camera is being used to ensure the robot's movement is always toward the desired object, then eventually the robot's fingers should collide with the object. To prevent collision, the robot could monitor the sensors in the ends of the fingers (now closed) and stop when an object is detected (which would place it only a few inches in front of the hand.

The robot could then open the hand and confirm that neither of the sensors in the ends of the fingers are triggered. If one of the finger sensors is triggered, the robot could rotate slightly to that side, to ensure that the hand is centered in front of the object. Then, the robot could move forward until the horizontal IR sensor (mentioned earlier) detects that something is between the fingers. The hand can then be closed until the snap-action switch (also mentioned earlier) indicates contact has been made with the object, then a little more to ensure a good grip. If desired, a pressure sensor could be mounted on one of the fingers, and read with one of the Arm Expansion's analog inputs. This would allow better control over the hand's grip pressure.

At this point the arm can pick up the object and the robot can move to another location (perhaps bring the object to you, or place it in a basket).

As you can see, using the arm can be complicated, but if you break it down, the complete behavior is composed of many, relatively simple, smaller tasks. If your robot has appropriate sensors you can create a library of functions, each of which performs one of these simple, smaller, tasks. One basic behavior, for example, might close the robot's hand until it feels the object with the snap-action switch. Once you have the library of basic behaviors, more complex actions can be handled by simply combining the basic behaviors based on input from the sensors.

All this might seem like a lot of work to get your robot to pick up an object but if you think about it, the scenario just described demonstrates far more intelligence than you typically see robots doing today. Remember, there were three objects on the floor. You could have *verbally* told your robot to pick up the RED one. It identified the object, found its way to it, manipulated its hand to grasp it, and then picked it up – then it could even ask you what to do with it.

Not a Beginner's Project

Of course, this is not a beginner's project, but it does demonstrate what can be done if adequate sensory data is available. It also provides a general overview of the type programming necessary to create an intelligent machine.

We realize that most readers will probably not build a robot arm, so in Chapter 16 we will examine *in detail* all the programming needed to fully implement a less-complicated, goal-oriented, non-trivial, behavior that can be carried out on almost any RROS based robot. It is our hope that such an example will provide insights into the fundamental principles needed to build the robot of your dreams

Sample Code

Figure 15.14 shows a simple program that demonstrates some basic techniques for controlling the arm.

```
#Include "RROScommands.bas"
#Include "InitializationRoutines.bas"
gosub InitRROScommands
gosub InitRoboClawRobot

Main:
 rCommand(ExpansionSetup,ARM)
 delay 3000
 gosub InitArm
 PosData[2]=0
 gosub MoveArm
 gosub OpenHand
 PosData[0]=200
 PosData[1]=170
 gosub MoveArm
 gosub CloseHand
 PosData[0]=180
 PosData[1]=190
```

```
 gosub MoveArm
 PosData[0]=40
 gosub MoveArm
 PosData[0]=60
 PosData[1]=150
 gosub MoveArm
 gosub Drop
 gosub ParkArm
end

InitArm:
 //     shoulder elbow wrist rotate  o/c
 data ParkData; 250,  250,  240,   0,   0
 data MinData; 130,   50,   50,   0,   0
 data MaxData;  70,   80,   80,  70,  80
 data PosData; 250,  250,  240,   0,   0
 rCommand(SetServoIndex,0)
 for i=0 to 4
  rCommand(SetServoMin,MinData[i])
 next
 rCommand(SetServoIndex,0)
 for i=0 to 4
  rCommand(SetServoMax,MaxData[i])
 next
 rCommand(SetServoIndex,0)
 for i=0 to 4
  rCommand(SetServoPosition,ParkData[i])
 next
 // assuming default speed of 1 for each joint
 rCommand(EnableServos,2)
return

ParkArm:
 rCommand(EnableServos,1)
 rCommand(SetServoIndex,0)
 for i=0 to 4 // no wrist
  rCommand(SetServoPosition,ParkData[i])
 next
 delay 6000
 rCommand(EnableServos,0)
return

MoveArm:
 rCommand(EnableServos,1)
 rCommand(SetServoIndex,0)
 for i=0 to 2 // just shoulder,elbow,wrist
  rCommand(SetServoPosition,PosData[i])
 next
 delay 4000
return

OpenHand:
```

125

```
 rCommand(EnableServos,1)
 rCommand(SetServoIndex,4)
 rCommand(SetServoPosition,200)
 delay 3000
return

CloseHand:
 rCommand(EnableServos,1)
 rCommand(SetServoIndex,4)
 rCommand(SetServoPosition,20)
 delay 3000
return

Drop:
 rCommand(EnableServos,1)
 rCommand(SetServoIndex,4)
 rCommand(SetServoPosition,150)
 delay 3000
return
```

Figure 15.14: This sample code shows some basic techniques for controlling the arm.

Even More Prototypes

We have added a couple new prototypes since the RROS was first introduced. The robot in Figure 15.15 is the steerable robot discussed first in Chapter 7. In its current form it has two digital perimeter sensors (far left and diagonal left) and a long range IR sensor mounted on the front. This lets it perform the wall following discussed in Chapter 16.

Figure 15.16 shows the initialization subroutine we used for our steerable robot. You should adjust the parameters as needed for your robot.

Figure 15.15: Our steerable prototype now has two digital perimeter sensors and a long range IR ranger on its front.

```
InitSteerableRobot:
 rCommport 44
 delay 100
 SetTimeOut 30000
 rlocate 0,0
 rCommand(MotorSetup,STEERABLE)
 rCommand(SensorSetup,DIGITAL+IRLONG)
 rCommand(CalibServoDrive,0)
 rCommand(EnableCounters,0)
 rCommand(SetSpeed, 60)
 rCommand(SetSlowDownSpeed,40)
 rCommand(SetMotorRamp,10)
 rCommand(SetDriveServoDir,1)
 rCommand(SetDriveServoWidth,50)
 rCommand(SetSteerServoDir,0)
return
```

Figure 15.16: This is the initialization routine for our steerable robot.

Another updated prototype is shown in Figure 15.17. It is built on the Rover 5 platform mentioned earlier in this chapter. We are now using it to test the SONIC ranging sensors available from our web page. Because of their low cost, they offer an alternative to Ping and Maxbotics sensors. They do require a totally different interface that utilizes one more pin that Pings. For that reason, if you use a rear bump sensor it must be digital.

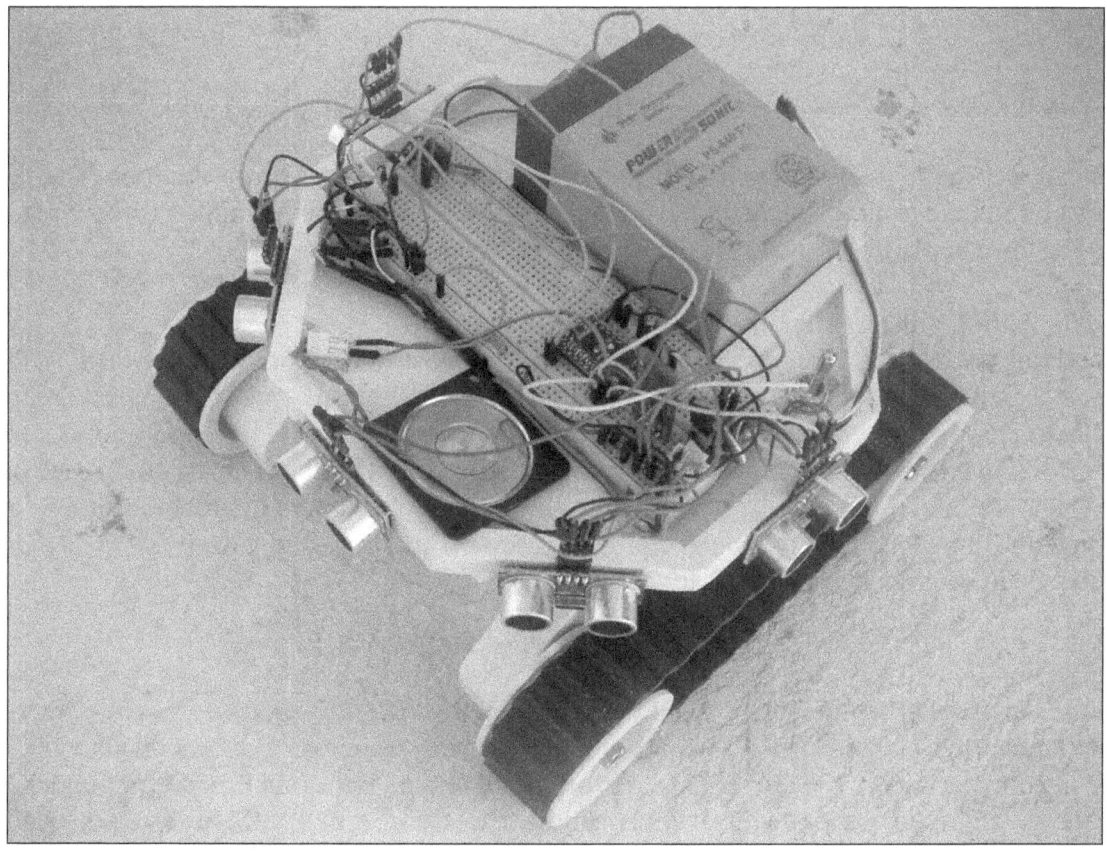

Figure 15.17: The Rover 5 has become the prototype for our SONIC ranging sensors.

Figure 15.18 shows the initialization routine we used for our Rover 5 prototype.

```
InitRover5:
 rCommport 47
 SetTimeOut 50000
 rlocate 0,0 // 255,0
 rCommand(MotorSetup,SMALLDC+ENCODERS)
 rCommand (SensorSetup,SR04+FIVERANGE) //+HMC6352 if you have a compass
 rCommand(SetMotorRamp,5)
 rCommand(SetClicksPerDiam,200)
 rCommand(SetClicksPer90,185)
// note: tanks do not make accurate turns
 rCommand(SetReducForwRight,0)
 rCommand(SetReducForwLeft,0)
 rCommand(SetReducBackRight,0)
 rCommand(SetReducBackLeft,0)
 rCommand(SetRotationTime,45)
 rCommand(SetMoveTime,14)
 rCommand(SetSpeed,65)
 rCommand(SetSlowDownSpeed,50)
 rCommand(SetSlowDown2,40)
 rCommand(SetCCdivisor,1) // need bigger
        //correction for treads
 rCommand(SetBumpDist,10)
 rCommand(SetProxDist,15)
 rCommand(EnableCounters,1)
 rCommand(SetRobotAngle,90/2)
 rCommand(SetTurnStyle,100)
 rCommand(SetRROStimeout,15)
return
```

Figure 15.18: This initialization routine is for our Rover 5 robot which utilizes low-cost SONIC ranging sensors.

As you can see, initially telling the RROS everything it needs to know about the robot's sensors and motors can take many commands – which is why we added special commands for initializing the RB-9 and Arlo platforms as discussed in Chapter 11.

Arlo: The Robot You've Always Wanted
After building many prototypes and writing many magazine articles on how to use the RROS to build powerful robots, we expected someone to create the kind of robot most hobbyists dream of owning. After waiting awhile, we finally decided to do it ourselves. Figure 15.19 shows a picture of our Life-Sized Arlo robot featured in a 4-part series starting with the January 2015 issue of Servo Magazine. A book titled *Arlo: The Robot You've Always Wanted* (available from Amazon.com summer 2015) gives detailed information on building your own Arlo. You can see a YouTube video of Arlo in action with this link.

http://youtu.be/ohpLRN-y2wY

Arlo has far more sensors than other robots. He has 6 ultrasonic ranging sensors around his base that implement 9 virtual sensors. There is a compass, battery monitoring, a beacon detector, 3 line sensors, 3 cameras, and two more ranging sensors (both IR and ultrasonic) on a turret under his head. He uses Parallax motors with wheel encoders. He has two arms with 4 sensors on each gripper. He can accept voice commands and respond verbally with an animated graphical face. He really is the robot you've always wanted.

Figure 15.19: Arlo: The Robot You've Always Wanted

Chapter 16

RROS Programming Examples

In the last chapter we provided an overview of the prototypes used to test the RROS chip during the development process. That chapter also provided an initialization routine for each of the prototypes to demonstrate how to configure various options. This chapter will assume that you have prepared an initialization routine for your robot and saved it under the name InitMyRobot.

In most cases, the example programs in this chapter will run on any RROS based robot. The exception to this is the steerable robot which is inherently different from our simulator-based prototypes. Many motor commands and sensor readings can be used with the steerable robot, but the nature of its movements will often require alternative algorithms when developing a specific behavior.

Testing the Sensors

The program in Figure 16.1 can be used to test nearly all the sensors on your robot. Set the value of the variable Real to True or False depending on whether you are testing a real robot or the simulator.

In the simulator mode, moving the mouse will move a yellow ball around the screen that can be used to trigger sensors. Move it close to the robot, for example, and you will see how the rBumper and rFeel sensors are triggered. Moving the ball in front of the robot will trigger the beacon sensor and the ranging sensor. You can even trigger the line sensors (rSense) by moving the ball close to the front of the robot. Use the right and left mouse buttons to rotate the robot, letting you see the compass readings change.

```
#include InitMyRobot
Real=False // set to true to use REAL robot
BeaconNum = Yellow // beacon color (or number) to detect
// Initialize the simulator OR the real robot
if Real
  // gosub appropriate init routine
  // for YOUR robot
  gosub InitMyRobot
else
  rLocate 400,300
  x=10
  y=10
  d=0
endif

xyString 1,480,"Bumpers      Feel        Line"
xyString 335,480, "    Range(front)   Compass"
xyString 720,480,"Beacon"
while(1)
 xyString 20,500,rBumper();"";rFeel();"";rSense();"";\
```

```
  rRange(0);"";rCompass();" ";rBeacon(BeaconNum)," "
 rForward 0
 if(!Real)
  rLocate 400,300,d
  rInvisible BeaconNum
  call DrawCircle(x,y,20,white,white)
  ReadMouse x,y,b
  call DrawCircle(x,y,20,black,BeaconNum)
  if b=1
   d--
   if Real then rTurn 5
  endif
  if b=2
   d++
   if Real then rTurn -5
  endif
 endif
wend

sub DrawCircle(x,y,r,c,cc)
 LineWidth 3
 Circle x-r,y-r,x+r,y+r,c,cc
return
```

Figure 16.1: This program can test your robot's sensors

The program in Figure 16.1 does use a little trickery. In order to prevent collisions when the ball is moved too close to the simulated robot, the program constantly re-locates the robot within its environment – something you generally would not want or need to do. There are also a few other aspects of the program that deserve mentioning.

First, the rForward 0 command is not needed in the program as it is written. If you modified the program though, to only test the bump, feel, and line sensors, then it would be needed. The reason for this is that these sensors are so time sensitive that the RobotBASIC does NOT require a command to acquire this data (see the RobotBASIC HELP file for more details). Instead, the RROS reads and returns the data when almost any other command (such as rForward, rCompass, rBeacon, etc) is executed. This means that the most recent data from these sensors is typically already in RobotBASIC's memory whenever the robot is being controlled and can be retrieved *immediately* without having to take the time to interrogate the robot for it. This may sound complicated but it works very well, and in fact is necessary to obtain acceptable response times for these sensors. What is important is that your program cannot simply sit in a loop testing the value of one of these sensors without issuing some other robot command – something you would almost never do unless you were simply testing the sensors as we are in this program.

Notice that the program in Figure 16.1 assumes you have created and saved your own InitMyRobot routine as described in Chapter 15. If your init routine properly initializes your robot then the values of its sensors will be displayed instead of those of the simulator. Move your hand around your robot's perimeter and notice the sensor reading it produces. If you have a beacon, set it to 14 (Yellow) and confirm that it can be detected. Pressing the mouse buttons will even rotate your robot right and left so that you can view its compass readings. Of course, if your robot does not have some of the sensors being displayed, their values will be zero.

Testing Behaviors

The previous program demonstrated how to test the sensors, but it did not examine the process of using the simulator to develop a robotic behavior. In order to demonstrate such a process we need an example application that is complex enough to be meaningful, yet simple enough that we can concentrate on the principles involved. Our end goal will be to build a library of routines that allows the robot to move through a room to a beacon, even if there are objects blocking its path.

As we build the library of routines, we will examine numerous principles including:

- developing behavioral algorithms on the simulator
- converting from simulation to real-world
- improving the real-world performance
- combining behaviors to create applications

Designing Behaviors

We will start by developing an algorithm (a step-by-step plan) for accomplishing our goal. We can summarize the plan as follows.

> *The robot should rotate until it sees the designated beacon, then move toward it until it either arrives at the destination or is blocked by an obstacle. The robot should then follow the contour of the blocking object (a wall-hugging behavior) for a SMALL, possibly RANDOM, period of time These actions are repeated over and over until the beacon is found. Let's see what this actually translates to in terms of physical behavior.*

It is assumed that the beacon is mounted ABOVE obstacles that might be encountered, so the robot is always able to see the beacon, making it easy to create a FaceBeacon behavior. ForwardTillBlocked is also an easily created behavior that terminates whenever perimeter sensors detect a blocking obstacle. Going around the obstacle is a little more complicated because the robot does NOT know how big the obstacle is and therefore it does not know how long it must continue to move around the obstacle before starting over (rotating to face the beacon again).

This simply means that the robot should move around the object for a short distance, stop, and look for the beacon again. If the robot has not moved far enough around the object, then when it starts to move forward it will again encounter the obstacle and follow around it a little more before turning to face the beacon again. This means the robot might well stop several times to look for the beacon as it transverses around the object.

You might ask why we don't simple make the robot move along the contour for a longer period of time. Perhaps the object encountered is small, and the robot moves all the way around the object (back to its original position). If such a situation occurred, the robot would never be able to find its way to the beacon. A reasonable solution is to make the robot move some random distance (around the object) so that it will (within some reasonable number of tries) stop at some point on the other side of the object, allowing it to continue toward the beacon unimpeded.

Of course, it might encounter another object before it arrives at the beacon, but since all of the actions are being repeated it will *automatically* handle the new object just as it did the first.

Pseudocode

When developing a programming algorithm, programmers often use pseudocode – an easy-to-follow English-like shorthand description of the program's actions (as apposed to the cryptic programming statements often associated with low-level programming languages). A pseudocode version of our algorithm might look like this.

> *Perform any necessary initialization.*
> *Repeat the following until the robot is at the beacon.*

Turn and face the beacon.
Move forward until blocked by an obstacle.
Go around the obstacle for a random distance.

If you study this example, it should be easy to see that if the robot can perform all the basic actions in the pseudocode, that it will eventually accomplish its goal no matter what kind or how many obstacles it encounters. One of the great things about RobotBASIC is that it has a very readable, English-like syntax, that is naturally very much like its own pseudocode. For example, the following RobotBASIC program fragment describes the same actions as the previous pseudocode.

```
gosub Initialization
repeat
  gosub FaceBeacon
  gosub ForwardTillBlocked
  gosub GoAround
until Done
```

You might be thinking that writing a program to find a beacon in a cluttered environment cannot be this simple, and you are right. We still have to create the code for each of the subroutines that implement the individual behaviors like facing the beacon. Creating the sub-behaviors is much easier than attacking the larger overall problem though, so it is probably easier than you might think. Let's start by creating the FaceBeacon behavior.

For now, we will develop the behavior solutions using the RobotBASIC simulator because it is far easier than trying to create, test, and debug similar code using a real robot. More on this later.

The following code shows a simple subroutine that will cause the robot to face a yellow beacon (beacon #14).

```
FaceBeacon:
  while not rBeacon(Yellow)
    rTurn 1
  wend
return
```

Yes, it is just that easy. As long as the robot does NOT see the yellow beacon, it turns to the right. Once the beacon is seen, the loop ends and so does the subroutine. Now let's look at some simplified code for ForwardTillBlocked. **Note:** As mentioned in previous chapters, there is a special rCommand to allow real robots to find beacons more efficiently.

```
ForwardTillBlocked
  While not (rFeel()&14)
    rForward 1
    if rSense(Yellow) then end
  wend
  Done = True
return
```

The rFeel() function reads the five proximity sensors on the simulated robot. By ANDing the sensor reading (using the & operator) with 14 (binary 01110) we restrict the reading to only the middle three sensors, making the robot only detect objects directly in front if it. If an object is NOT detected, the robot moves forward. When an object is detected, the loop, and thus the subroutine ends. Notice that the rSense() sensors (line sensors) are also checked to see if the robot has reached the beacon so that the program can be terminated that if it has.

Combining the Modules

Below is an Initialization module that places the beacon in the upper right corner of the screen and locates the robot and a blocking obstacle in random positions. It also sets up invisible colors so that the robot will interpret the beacon color properly, so the robot can leave a trail as it moves. Comment out the gosub GoAround statement in the main program module since we have not yet created that subroutine and then combine it with the other modules to create a single program. Each time you run the program the robot will either move to the beacon and stop, or move toward the beacon until it is blocked and stop, as shown in Figure 16.2.

```
Initialization:
 circle 750,50,790,10,yellow,yellow
 rLocate 50+Random(300),50+Random(550)
 x=400+Random(250)
 y=50+Random(550)
 r=50+Random(50)
 circle x-r,y-r,x+r,y+r,Red,Red
 rInvisible green,yellow
 LineWidth 2
 rPen down
return
```

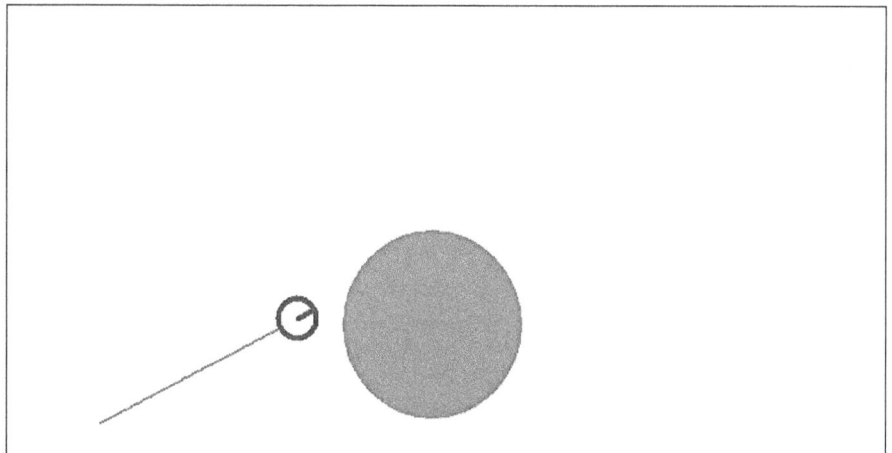

Figure 16.2: The robot will move to the beacon unless it is blocked.

As you can see from Figure 16.2, the robot will face the beacon, move toward it unless it is stopped. In this program, we have assumed that the robot will move in a straight line, which the simulator typically does. A real robot should certainly do this reasonably well if your robot uses wheel encoders or if you have calibrated its movements. Even so, it is possible for one of the wheels to slip on the floor (which still registers counts on the encoders) allowing the robot to drift from its intended course. Since our real robot must contend with problems like this, it would be nice if the simulation was subject to the same type of errors. If we add this line just after the rPen down statement,

rSlip 20

then the simulated robot will act more like a real robot by generating up to 20% error (far more than a real robot), as you can see by Figure 16.3.

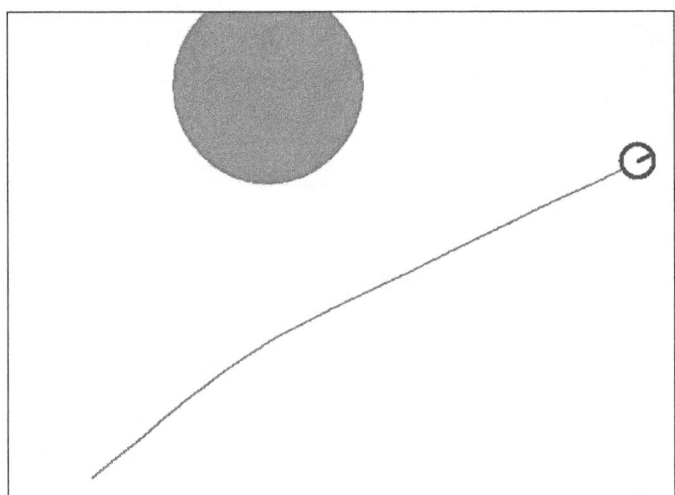

Figure 16.3: The simulator can create random error in order to better simulate a real robot.

Our program can correct for this drift if we modify the ForwardTillBlocked routine so that it monitors the beacons signal and stays on course. The modification is shown in the new module below. Notice how the robot turns to the right when it sees the beacon and to the left when it does not. This causes the robot to constantly adjust to maintain its course toward the beacon.

```
ForwardTillBlocked:
  While not (rFeel()&14)
   rForward 1
   if rSense(Yellow) then end
   if rBeacon(Yellow)
    rTurn 1
   else
    rTurn -1
   endif
  wend
  Done=true
return
```

The modified program now performs as it should even with the random slip. It faces the beacon and moves toward it until it finds the beacon, or until it is blocked by an obstacle. In order to complete our program, we need a way for the robot to follow along the contour of the object blocking its path until it gets to the other side of the obstacle. At that point, the robot can turn to face the beacon again and resume moving toward the beacon, thus completing the task. This means we must develop an algorithm to provide our robot with the ability to follow the contour of an object. Such a behavior can be quite complicated if it must work correctly with objects that have complex shapes, but to keep things simple we will assume the blocking objects are only modestly complicated.

Wall Hugging

A robot's ability to stay close to a wall, to follow its contour can be valuable in many ways. For example, it can be a basic movement when solving a maze or finding a way around an obstacle that blocks the robot's path. A *perfect* wall-hugging behavior probably would use multiple types of sensors (perhaps rFeel, rBumper, and rRange), but in order to reduce the complexity of this example (as mentioned earlier), we will not try to make it absolutely perfect.

Because we are trying to keep things simple, our goal will be to create an algorithm for following the contours of a wall using *only* the rFeel sensors. The basic idea is relatively easy to conceive. Figure 16.4 shows two orientations of a robot following along a wall. When the robot is in position A, where only the left-side sensor sees the wall, the robot should turn to its left, bringing it closer to the wall, as shown in position B of the figure. When both the left sensor and the left diagonal sensor sees the wall, the robot should turn right, away from the wall.

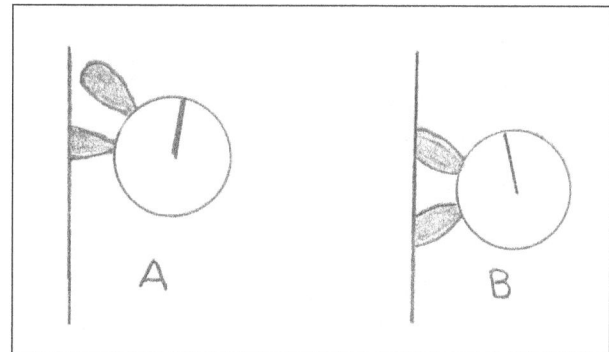

Figure 16.4: Monitoring the two proximity sensors shown in this diagram can be the basis for a wall following behavior.

Using the principles of Figure 16.4, we can create a subroutine that can follow a wall as shown below. Note: Some sensors (such as the top sensor shown in Figure 16.4A) might not be able to detect the wall properly, especially if the wall is very smooth so that light or sound will angle off away from the robot instead of reflecting back to the robot. For such situation a different algorithm needs to be developed. Try to develop a different technique using the simulator or compare your ideas to various ones examined in some of our books and magazine articles.

```
FollowWall:
 rTurn 60    // turn to align with wall
 while 1
  d=rFeel()  // read proximity sensors
  if d=16   // left sensor
   rTurn –1 // turn left
  elseif d=24 // left and diagonal sensors
   rturn 1  // turn right
  endif
  rForward 1
```

```
 wend
return
```

Testing the Algorithm

All of the above principles can easily be implemented in the RobotBASIC simulator as shown in Figure 16.5. The program uses three subroutines to perform the major tasks. One draws an object with both concaved and convexed walls to be hugged. Another moves the robot until it bumps into the wall.

The final subroutine performs the actions described earlier for following the wall. It starts by turning to the right 60° as an attempt to approximately align itself with the wall to be followed. The robot is constantly moving forward, and turns left if the side sensors is NOT seen, and right when the diagonal sensor IS seen. Since the robot has been programmed to leave a trail, its actions will look like that of Figure 16.6.

```
main:
 gosub DrawWall
 rLocate 400,500
 rInvisible green
 LineWidth 2
 rPen down
 gosub FindWall
 gosub FollowWall
end

DrawWall:
 circle 300,300,600,400,blue,blue
 circle 350,250,550,400,blue,blue
 circle 350,350,550,450,white,white
return

FindWall:
 while not(rFeel()&13)
  rForward 1
 wend
return

FollowWall:
 rTurn 60     // turn to align with wall
 while 1
  d=rFeel()  // read proximity sensors
  if d=16    // left sensor
   rTurn -1 // turn left
  elseif d=24 // left and diagonal sensors
   rturn 1  // turn right
  endif
  rForward 1
 wend
return
```

Figure 16.5: This program shows the basic principles for following a wall.

As you can see from Figure 16.6, the plan we formulated for the wall-following behavior works reasonably well, as long as the wall is composed of modest curves. When an abrupt change is encountered though, the robot fails. The reason for this can be seen in Figure 16.7. As you can see from the figure, we must use the front sensor instead of the two left-side sensors to detect the type of obstruction that caused the problem for our simulation.

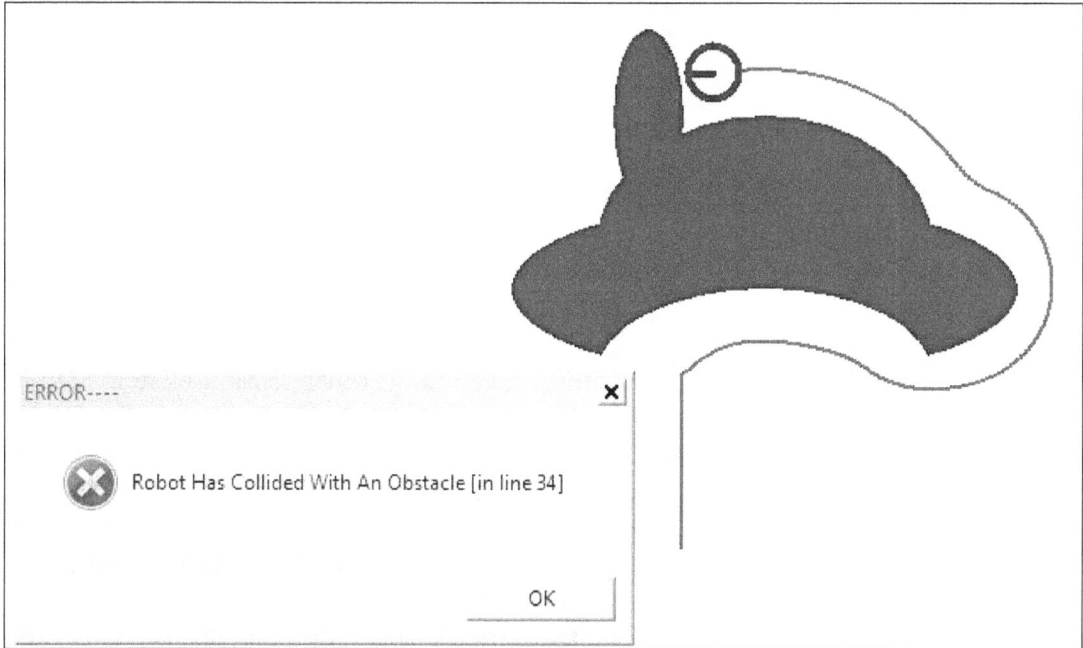

Figure 16.6: This output is produced when the program in Figure 16.5 is executed.

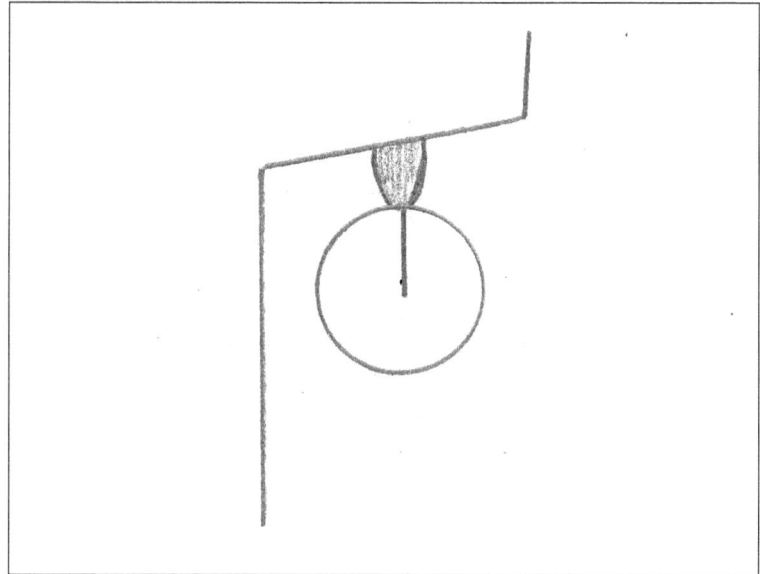

Figure 16.7: The robot needs to use its center sensor to detect abrupt changes in the objects contour.

When the front, center sensor (as shown in Figure 16.7) sees an object the robot must turn to the right. We can modify the FollowWall subroutine to perform this action as shown in Figure 16.8.

```
FollowWall:
 rTurn 60
 while 1
  d=rFeel()
```

```
  if d&4      // check for front sensor
   rTurn 1    // turn to the right (change to 45)
  elseif d=16
   rTurn –1
  elseif d=24
   rturn 1
  endif
  rForward 1
 wend
return
```

Figure 16.8: This modification to the FollowWall checks the front
sensor and turns the robot to the right when something is detected.

Notice in Figure 16.8, that the sensor reading (d) was ANDed with 4 (using the & operator). This allowed the front sensor to be detected no matter what state the other sensors were in. If we had tested the condition d=4, for example, all the other sensors would have to be zero for the condition to be true. If you are not familiar with this type of programming operation then it is recommended that you read a book on programming principles such as our *Robot Programmer's Bonanza*.

If you alter the program as described above though, it still crashes. Looking back at Figure 16.6, it is easy to see why. When the robot detects an obstacle directly in front of it, it needs to turn QUICKLY to the right in order to avoid a collision. If we change the associated rTurn 1 to rTurn 45, the robot will indeed make it though the problem area, but then creates a new problem as shown in Figure 16.7.

With a little thought it is easy to see why the new problem occurs. If the robot finds itself too far away from the wall, then NONE of the sensors are triggered. If we look back at our program we see that we have not written any code to handle such a situation – so the robot just moves forward as we can see in the figure.

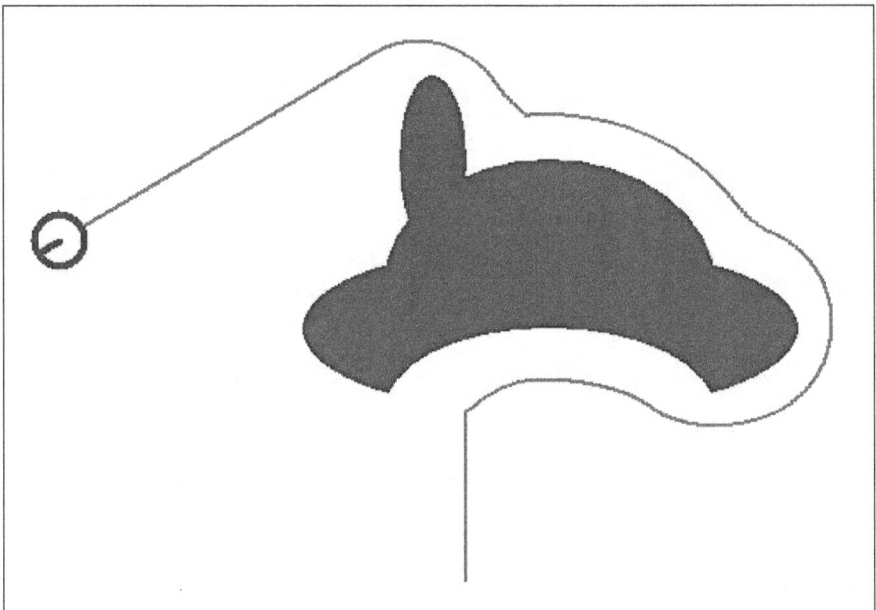

Figure 16.7: The robot gets a little further along the wall, but still fails.

Figure 16.8 shows a new FollowWall that can detect when no sensors are triggered and turn back to the left.

140

```
FollowWall:
 rTurn 60
 while 1
  d=rFeel()
  if d&4
   rTurn 45
  elseif d=16
   rTurn -1
  elseif d=24
   rturn 1
  elseif d=0  // checks for NO sensors
   rTurn -1  // change to 15
  endif
  rForward 1
 wend
return
```

Figure 16.8: This FollowWall routine is still not perfect.

If you run the program with the changes shown in Figure 16.8, you will get the output shown in Figure 16.9.

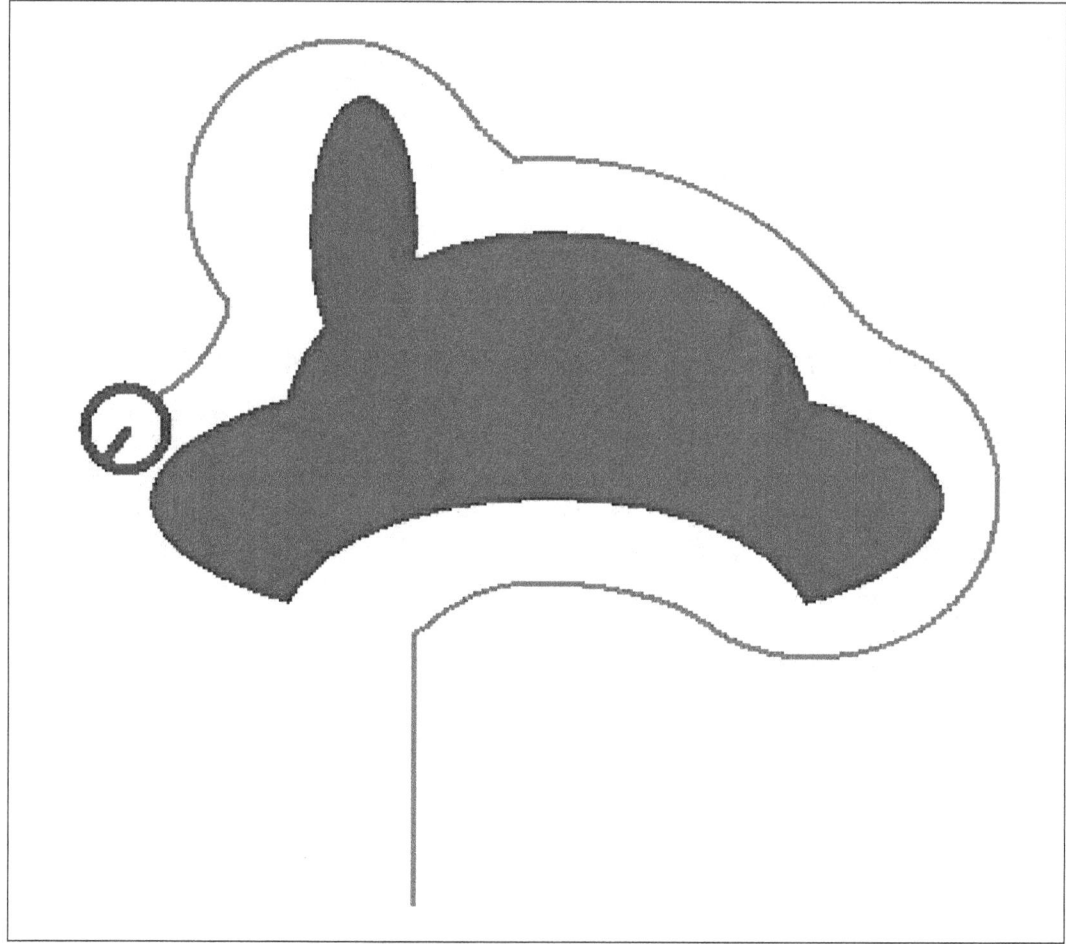

Figure 16.9: The robot moves better, but it still strays a bit too far from the wall after rounding the protrusion, and, it still eventually crashes.

With the new changes the robot makes its way around the protrusion, but if you wait long enough, eventually it still crashes. This problem was a little harder to track down, but the basis for it is that, in certain situations, the robot waits too long before turning away from the wall. Figure 16.10 shows how this can happen.

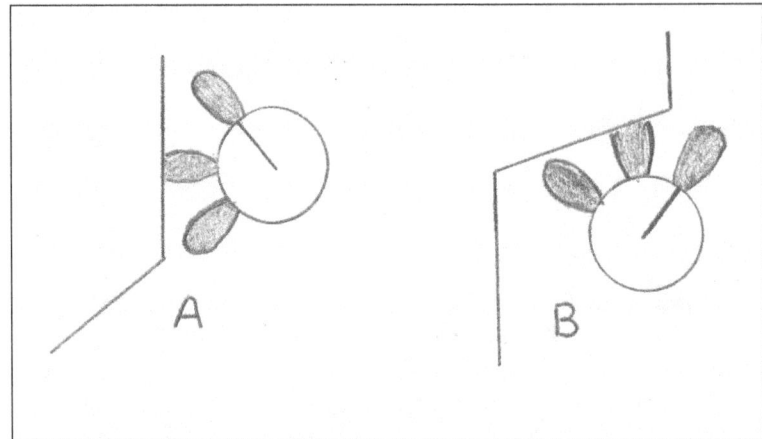

Figure 16.10: Although it is unusual, it is possible for only the diagonal sensor to be triggered.

As shown in Figure 16.10, it is possible for the robot to be approaching the wall too closely without the left and front sensors triggering (only the diagonal sensor is triggered). Since our algorithm only turns to the right based on readings from the left and front sensors, the robot must wait until one of those sensors is triggered before turning away. You can see from the figure though, that the robot can be warned slightly earlier if we check the diagonal sensor. Just checking it is not enough, though.

Remember, we ALREADY check the diagonal sensor (the condition d=24) and turn away from the wall. This test though, checks to see that BOTH the left and the diagonal sensor are triggered simultaneously. The situation described by Figure 16.10 has ONLY the diagonal sensor being triggered. When both the left and diagonal sensors are triggered we only need a gentle correction, so we turn away slowly, but when the diagonal sensor is triggered alone, it makes more sense to turn away at a faster rate (but perhaps not as fast as when no sensors are triggered).

There is actually another time when it also makes sense for the robot to turn a little faster. Refer back to Figure 16.9 and notice that the robot loops further from the wall than its normal movement, right after it rounds the protrusion at the top of the obstacle. This situation is occurring when the robot does not see any sensors so we can correct it by having the robot turn back a little faster when d=0.

Figure 16.11 shows our final WallFollow routine and Figure 16.12 shows the output when the program is run.

```
FollowWall:
 rTurn 60
 while 1
  d=rFeel()
  if d&4
   rTurn 45  // fast turn
  elseif d=16
   rTurn -1  // slow turn
  elseif d=24
   rturn 1   // slow turn
  elseif d=0
   rTurn -15 // medium turn
```

```
  elseif d=8
   rTurn 15  // medium turn
  endif
  rforward 1
 wend
return
```

Figure 16.11: In this version, the robot turns at different rates for different situations.

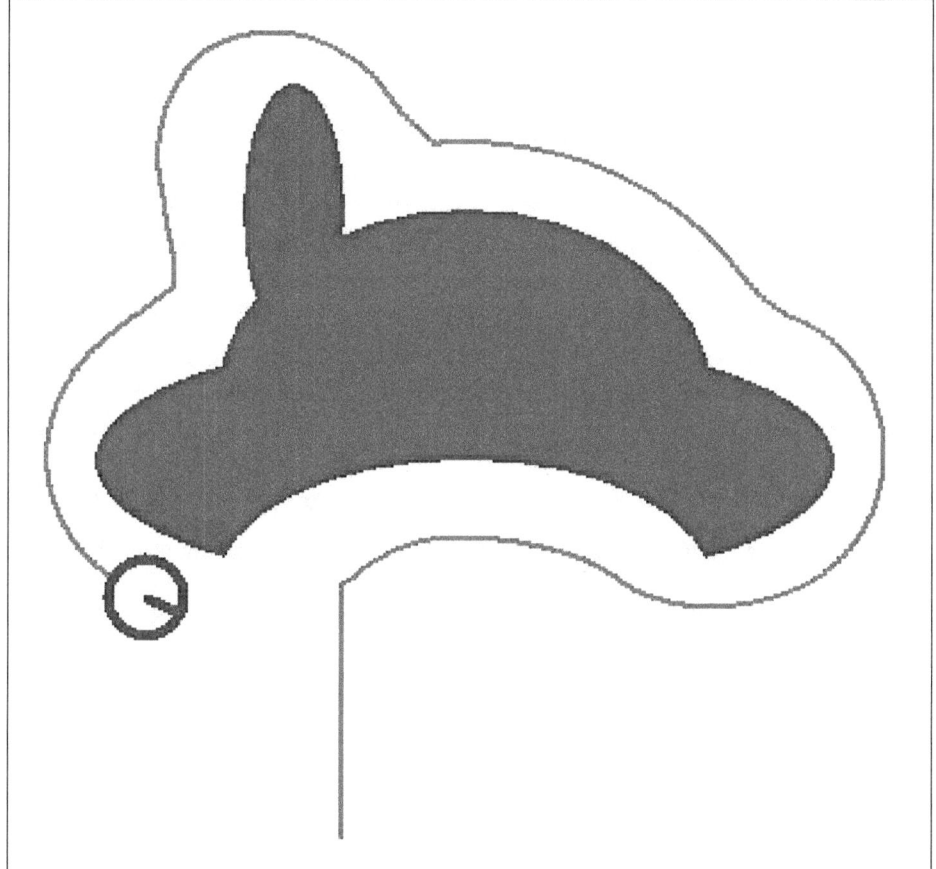

Figure 16.12: The robot now hugs the wall closely without collision.

As a robot programmer it is your responsibility to anticipate every possible problem that the robot might encounter so that your robot can perform properly no matter what situation it finds itself. The robot simulator can be a big help with anticipating potential problems because it allows you to see things that are often difficult to view with a real robot.

Better Algorithms

It is important to remember that this entire behavior was created using only the proximity (rfeel) sensors. If the algorithm also check the rBumper and rRange sensors the robot could check for additional conditions and anticipate much quicker – allowing it to respond more appropriately to an even more diverse environment.

Working with the simulator can certainly be productive, but everything we learn must eventually be applied to a real robot, and we will explore that option shortly. For now though, let's continue on with our goal of using our wall-following behavior to allow our robot to find its way to a beacon.

The problem we have with our current FollowWall routine is that it follows the wall *forever*. If you look back at our original ideas for a solution, we needed the robot to follow along the wall for a random amount of time, then turn to face the beacon, etc. We can easily place most of our original code inside a for-loop (instead of a while) so that the robot can continue its actions for a limited amount of time. It would also be nice to be able to easily specify the amount of time desired.

In addition to the standard BASIC subroutines, RobotBASIC also provides a more modern function-like routine that allows the passing of variables. This new style also utilizes local variables making it easier to create reusable library functions.

Figure 16.13 shows a complete program that places the robot in a random position and lets it find its way around two obstacles to a beacon. This new program implements all of the ideas previously discussed, but with the new function-style subroutines. We will discuss them briefly here, but if they are totally new to you, refer to the RobotBASIC HELP file.

Look first at the new FollowWall routine. It is created using the sub command. Notice also that the function name is followed by a variable in parenthesis. When you call the function (see the main module) we specified 300 as the parameter to be passed. This simply means that the variable HowFar will have that value inside the function itself.

The FaceBeacon routine is also implemented as a function so we can pass it the beacon color (number). The MoveToBeacon is a little different. First of all, we pass it two parameters the first of which is the color of the beacon, just as with the FaceBeacon routine. The name of the second variable though, is preceded by an &. This tells RobotBASIC that the value of the variable is not being passed. Instead, the called routine should use the actual variable listed in the original call statement. In this example, it simply means that when the code inside MoveToBeacon references the variable f, it is actually using the variable found.

Perhaps a couple of examples can further illustrate this. A statement such as x=f, for example, would actual make x equal to the value of found. A statement such as f=4, would actually make found equal to 4. This is an easy way for functions to pass information *back* to the calling module. In this case, the MoveToBeacon routine always sets the value of the variable found to true or false in order to let the main module know whether it actually found the beacon or not.

```
main:
 gosub DrawWall
 rLocate 20+random(300),550
 rInvisible green,yellow
 LineWidth 2
 rPen down
  repeat
  call FaceBeacon(yellow)
  call MoveToBeacon(yellow,found)
  if !found then call FollowWall(300)
 until found
end

DrawWall:
 circle 300,300,600,400,blue,blue
 circle 350,250,550,400,blue,blue
 circle 360,200,400,300,blue,blue
```

```
 circle 350,350,550,450,white,white
 circle 120,450,325,500,blue,blue
 circle 700,100,750,150,yellow,yellow
return

sub FollowWall(HowFar)
 rTurn 60
 for i=1 to HowFar
  d=rFeel()
  if d&4
   rTurn 45  // fast turn
  elseif d=16
   rTurn -1  // slow turn
  elseif d=24
   rturn 1   // slow turn
  elseif d=0
   rTurn -15 // medium turn
  elseif d=8
   rTurn 15  // medium turn
  endif
  rforward 1
 next
return


sub FaceBeacon(c)
 while !rBeacon(c)
  rTurn 1
 wend
return

sub MoveToBeacon(c,&f)
 while 1
  if rSense(yellow)
   f=true
   return
  endif
  if rFeel()&14
   f=false
   break;
  endif
  if rBeacon(c)
   rTurn 1
  else
   rTurn -1
  endif
  rForward 1
 wend
return
```

Figure 16.13: This program demonstrates all the principles discussed so
far in this chapter as it moves the robot around objects to find a beacon.

If you run the program in Figure 16.13, it will move the robot in a manner similar to that shown in Figure 16.14. The path will be slightly different each time the program is run because the robot is located randomly. Watch the robot as it moves and you will see it will periodically stop the wall-following behavior and look for the beacon. You can change how far it travels by altering the parameter in the call to FollowWall.

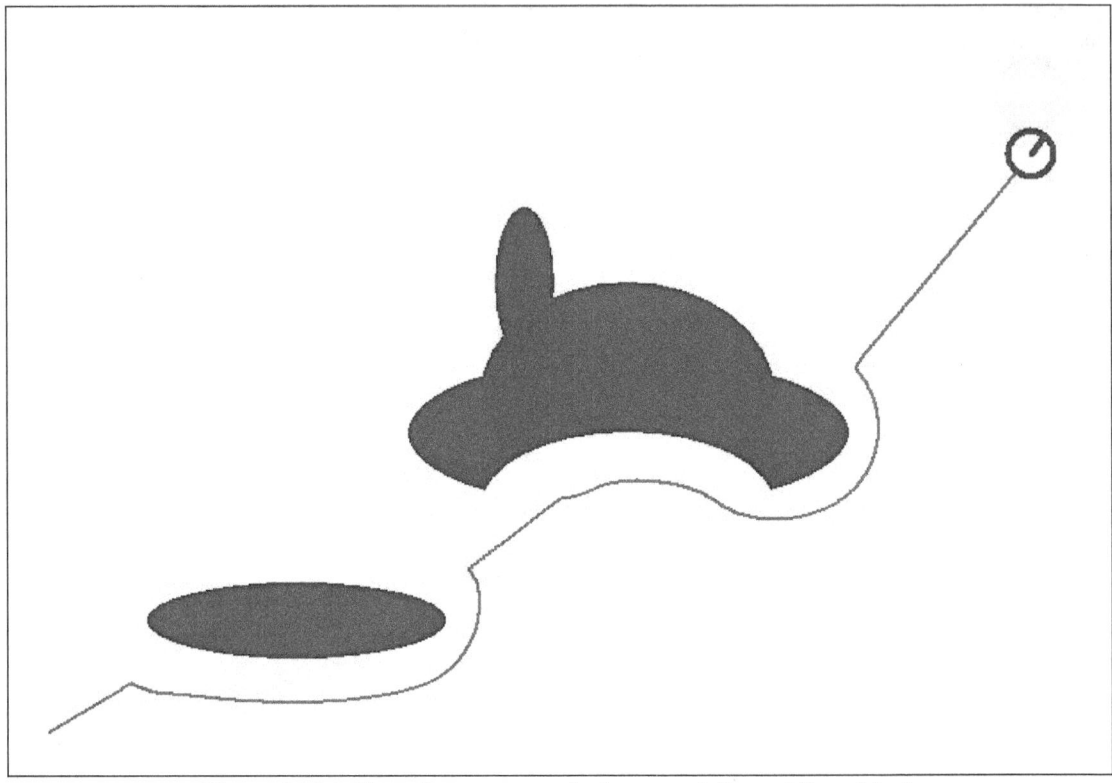

Figure 16.14: The robot moves to the beacon while avoiding objects
by going around them using a wall-following behavior.

Now that we have completed our work with the simulator, we are ready to make the program work
with a real robot.

Moving the Behavior to a REAL Robot

To keep things simple, let's assume the program we are converting is only a basic wall-follow
routine (as apposed to the more complex find-beacon application). **Note**: Even though we will be
using the basic wall-follow behavior, we will use the function-style subroutine when creating
FollowWall. We can create the main code that will work on both the simulator and a real robot as
shown in Figure 16.15. It starts by including the initialization routines for both the RROS and the
real robot to be used (as described though out the text and especially in Chapter 15).

Notice that a variable Real has been set to False, and then used to decide how to initialize the
robot to be used. Notice also that when it is set to True, that an initialization routine for the real
robot is called. We tested this routine with our small DC motor prototype (pictured in Chapter 15)
and the robot followed walls with modest contours just as the simulation did, with one exception.
The real robot moved with a noticeable (and annoying) jerky motion.

```
#include InitRROScommands
#include MyInitRoutine
main:
 gosub InitCommands
 Real = False
 if Real
  Gosub InitDCrobot
 else
  gosub DrawWall
  rLocate 400,500
```

```
 rInvisible green
 LineWidth 2
 rPen down
 endif
 gosub FindWall
 call FollowWall(1000)          (
end
```

Figure 16.15: This new Main module allows the same program
to control either the simulation or a real robot.

The reason for this is simple. The simulated robot makes its turns by ROTATING, which means one wheel turns backwards while the other moves forward. The jerky motion occurs when the real robot (which unlike the simulated robot, has to deal with mass and momentum) quickly reverses the direction of one of its wheels (as discussed elsewhere in this text). This jerky motion can be eliminated by altering the way the real robot turns.

In its default state, the real robot turns just like the simulation – it rotates around its center by turning one wheel forward and one wheel backward. If we just stopped one wheel instead of reversing it, the robot would still turn. The turn of course would be slightly slower, but since no part of the robot is now moving backwards, the jerk is eliminated. This new turn is certainly better for some situations (like wall following) but when maneuvering in tight confines, the rotational turn is still preferred.

Earlier in this book an rCommand was discussed that told the RROS what kind of turns to make. It took the form of:

rCommand(SetTurnStyle, param)

where param could be any number from 0 to 100. A value of 100 produces the normal rotational turn used by the simulator. All other numbers determines the speed of the slower wheel compared to the faster wheel. A param of 40, for example, will cause the slower wheel to run at 40% of the faster wheel's speed. The smaller the number, the faster the turn. Large numbers create more of a drift to one side rather than an actual turn.

Earlier in this chapter we saw that a robot follows a wall better when it makes turns at various rates based on the sensory conditions. Figure 16.16 shows a new FollowWall routine that works with both the simulator and a real robot.

```
sub FollowWall(HowFar)
 rTurn 60
 rCommand(_SetTurnStyle, _SlowTurn)
 for I=1 to HowFar
  d=rFeel()
  if d&4
   if _Real
    rCommand(_SetTurnStyle, _FastTurn)
    rTurn 1
    rCommand(_SetTurnStyle, _SlowTurn)
   else
    rTurn 45  // fast turn
   endif
  elseif d=16
   rTurn -1 // slow turn
  elseif d=24
   rturn 1  // slow turn
  elseif d=0
   if _Real
    rCommand(_SetTurnStyle, _MediumTurn)
```

```
      rTurn -1
      rCommand(_SetTurnStyle, _SlowTurn)
    else
      rTurn -15 // medium turn
    endif
  elseif d=8
    if _Real
      rCommand(_SetTurnStyle, _MediumTurn)
      rTurn 1
      rCommand(_SetTurnStyle, _SlowTurn)
    else
      rTurn 15  // medium turn
    endif
  endif
  if not Real then rForward 1
 next
return
```

Figure 16.16: This FollowWall routine will work
on both the simulator and a real robot.

There are several things about the routine in Figure 16.16 than deserve discussion. First, notice that most of the conditions did NOT need any modification because the normal turn style used throughout the routine is SlowTurn. When a medium or fast turn is needed, the turn style is changed accordingly, then changed back to the normal slow state.

Notice also that numerous variables are preceeded by an underscore character (_). When used this way inside a function style subroutine, RobotBASIC assumes the variable is NOT local to the function, but is one of the global variables associated with the main program. We could always pass the information associated with such variables, but this technique is an easy short-cut that most people love after using it.

Notice also that the rForward command is only used when the robot is simulated (not Real). It is necessary for the simulation, because the turns on the simulated robot only rotate – meaning the simulated robot NEVER moves forward unless told to do so.

With our real robot though, the slow turns do advance the robot, even if only slightly. If we also move the robot forward, it partially negates our slow turn behavior. For these reasons, we do not want the rForward with the real robot.

Controlling the Slow Turns
Of course, the main program or the initialization subroutine needs to establish values for the slow turn parameters. Those that worked well for our DC robot are shown below.

```
SlowTurn = 70
MediumTurn = 30
FastTurn = 10
```

Of course these parameters will vary based on the physical characteristics (type of motors, the robot's weight, speed being used, etc.) of the robot being used. The robot's weight and speed, for example, will alter the actual turn style because of the robot's momentum. When we ran this program with our servomotor robot and our man-sized robot we only had to make minor changes to the turn-style parameters above in order to get a proper wall-following behavior. NO OTHER changes were necessary. Since the value of these parameters are unique to each type of robot, we recommend that you initialize their values in your robot's initialization routine. That way, if you have several robots, the parameters they need will be established automatically. This means that some variables (for example: Real, SlowTurn, etc.) should be "reserved" for this special use.

Behavior Libraries

We recommend that you create function-style routines for the behaviors you develop. For example, you could have behaviors for FollowWall, FollowLine, FaceBeacon, GotoBeacon, ForwardTillBlocked, GotoXY, FindBatteryCharger, etc. You could also develop a complete set of routines for controlling the arm.

If properly developed, such routines could be used with almost any RobotBASIC robot. We encourage those that develop such routines to write articles for robot magazines such as *Servo* and/or send them to us so we can make them available for download to other RobotBASIC users.

Once you have a library of reusable behaviors, the time needed to create new applications will be greatly reduced.

Porting Algorithms to Steerable Robots

Many algorithms created for the simulator (and our 2-wheel robots) are not directly compatible with steerable robots. If you just look at the principles involved though it is usually easy to create appropriate algorithms. Figure 16.17, for example, shows a wall follow algorithm we created for our steerable prototype. For the most part it detects the same conditions and then makes the same kind of turns (slow, medium, fast) as our other robots – it just must make the turns by turning the front wheels instead of rotating.

Since a steerable robot cannot turn as fast as a rotating 2-wheel robot, we mounted a ranging sensor on the front of the robot (see Figure 15.5 in Chapter 15) in order to detect objects in front of the robot at a much longer distance than possible with standard perimeter sensors. This gives the robot the extra time it needs to turn away from objects in its path.

The only experimental changes that had to be made for the steerable robot to work properly was to fine-tune how much to turn for slow turns, medium turns and fast turns.

```
FollowWall: // for steerable robot
  while 1
  d=rFeel()&24 // mask off unused sensors
  if rRange()<12  // object in front
   rTurn 45    // fast right
  elseif d=8    // diagonal only
   rTurn 25    // medium turn right
  elseif d=16   // left sensor only
   rTurn -15   // slow left
  elseif d=24   // both left sensors
   rTurn 15    // slow turn right
  elseif (d&24)=0 // no senssors
   rTurn -40   // medium turn left
  endif
  rForward 1
 wend
return
```

Figure 16.17: The principles used in algorithms developed for the simulator
can be ported to other robots, such as a steerable robot.

Quick Reference
rCommand Parameters

The rCommands used to control many RROS functions were introduced throughout the book so that their relevance was shown in an appropriate context. This Appendix lists the commands in their numeric order so they can be used for reference.

It is worth noting although all the commands have been initialized to a variable name with the InitCommands subroutine (found in the RROScommands.bas #include file). While this generally does not produce a problem, for very large programs, it can *potentially* decrease the performance of some programs for the following reason. RobotBASIC is an interpreter, and that means it searches through a variable table in order to find the current value a variable. If the InitCommands subroutine is run early in a program, then all of the variables initialized there will be first in the table, and that means they will be searched every time a programs normal variables are accessed. Normally this will not cause any significant delay, but if your program needs to operate at absolute maximum performance, there are several options.

First, you can simply use the associated numbers inside rCommands instead of initializing variables to make your program easier to read. For example, you could use

rCommand(131,28)

instead of

rCommand(PlaySound, Phasor)

Another option would be to delay calling the InitCommands subroutine until you have used the main variables in your program (or at least the ones that will effect speed the most). This could be done by simply having a subroutine that sets all the variables in your program to 0.

Remember, in most situations the use of the InitCommands subroutine will not cause noticeable performance issues. We would be remise though, if we did not point out potential problems.

List of rCommands
The following is a list of the *commands* used within an rCommand along with their actual values and a short summary of the command's use.

110: CalibServoDrive
Enables calibration of the servo drive motors. If this items is set to FALSE the motors will NOT be pulsed when a speed of zero is active. A TRUE value causes the pulses to continue. The default condition is FALSE.

111: CalibrateCompass

Forces calibration of the currently active compass. Generally this means the robot will rotate for a short period (at the **SlowDownSpeed**) while readings are being taken. Usually the RobotBASIC timeout period must be increased to prevent error.

112: SetSenseInvMask

The mask parameter will be XORed with the rSense() data before it is returned allowing selective inversion of individual bits.

113: SetTurretOffset

The parameter provided adjusts the center position of the range turret. The default value is 128.

114: SetRotationTime

The parameter provided adjusts the rotation *time* for each degree specified in rTurn. Adjust so that rTurn 90 rotates the robot approximately 90°. Not used when wheel encoders are active.

115: SetMoveTime

The parameter provided adjusts the drive *time* for each increment of rForward. Adjust so that rForward 40 moves the robot forward a distance equal to its diameter.

116: SetLeftStopOffset
116: SetDriveStopOffset

The parameter provided calibrates the pulse width representing STOPPED condition for the left servo drive motor. Not used for DC motors. Two names are provided for clarity. Start with 128.

117: SetRightStopOffset
117: SetSteerStopOffset

The parameter provided calibrates the pulse width representing STOPPED condition for the right servo drive motor. Not used for DC motors. Two names are provided for clarity. Start with 128.

118: SetTurnStyle

The parameter provided sets the percentage of the standard speed that the slower motor will use during turns for rTurn 1 and rTurn -1. Other turns are not affected. Setting the parameter to 100 causes the two drive motors to run in opposite directions on all turns (which generally causes a jerky motion for many behaviors).

119: SetSpeed

The parameter provided becomes the standard speed. Values may range from 1-100.

120: SetCCdivisor

The parameter provided controls how the robot tries to correct itself when the two motors move at different speeds. The parameter will be divided into the standard speed to produce the slower speed used to the correct motor deemed to be moving too fast based on the wheel encoders. For example, a parameter of 4 means the faulty motor's speed will be decreased by 25%. The default value is 2.

121: SetMotorRamp

The parameter provided is added to or subtracted from the current motor speed as it ramps up to or down to the desired speed. In general, the larger the number the faster the robot will start and stop. The default value is 1.

122: SetBumpDist

The parameter provided sets the detection distance (in ½ inch increments) to be used by the virtual bumper (rBumper) sensors.

123: SetProxDist

The parameter provided sets the detection distance (in ½ inch increments) to be used by the virtual proximity (rFeel) sensors.

124: SetExpanTimeout

The parameter for this command controls how long the RROS will wait for a custom expansion to reply. Normally expansions should reply almost immediately, but if significant time is needed by an expansion, RROS must be told so that it can wait for an appropriate period before canceling the communication and assuming the data that should have been received is all zeros.

125: SetRobotAngle

The parameter provided is doubled and added to the angle obtained from the compass hardware. This allows the user to set the direction deemed to be due north (within 2 degrees). Doubling is required because the parameter is only 8 bits.

126: SetBeaconTime

Effectively sets the amount of time the robot rotates when looking for a beacon (FindBeacon). Set the parameter so the robot moves at least 360°.

127: SetClicksPerDiam

For use with wheel encoders only. Parameter specifies the number of wheel clicks needed to make the robot move forward a distance equal to its diameter.

128: SetClicksPer90

For use with wheel encoders only. Parameter specifies the number of wheel clicks needed to make the robot turn 90°.

129: Used for rPen

This is one of the auto-commands in RobotBASIC.

130: TurnToHalfAngle

Commands the robot to turn to a (room corrected) angle equal to twice the specified parameter. Doubling is required because the parameter is only 8 bits.

131: PlaySound =   131

Plays one of the parameter specified notes or sounds listed later in the Appendix. The lower six bits specifies the sound. The two MSBs specify the length.

132: TestCommand

This command is available for testing. The five bytes it returns include the specified parameter and 1, 2, 3, 4.

133: Reserved for Future Use

This commands is not currently available, but may be used in a future version of the RROS.

134: EnableMotors

The drive motors are always disabled when the RROS boots. They are automatically enabled when MotorSetup is executed. This command allows you to enable and disable the motors using a true/false parameter.

135: EnableCounters

The wheel-encoder-counters are automatically enabled for turns and drift compensation when encoders are specified with SensorSetup. This command allows you to enable and disable the use of the counters with a true/false parameter.

136: SetSlowDownSpeed

Allows a SlowDownSpeed to be specified. The SlowDownSpeed is used for turns and distances of more than 1 unit when Counters are *not* enabled. When counters are enabled, it is used to slow the robot down when a turn or distance destination is approaching.

137: MuteBeaconDet

A true/false parameter can be used to disallow or allow a sound to be emitted when the beacon detector is pointed at a beacon (audio feedback can be very useful when working with beacons).

138: MotorSetup

The three LSBs of the parameter provided chooses the types of motors to be used based on the following:

000	small DC driving by the RROS chip itself
001	continuous rotation servomotors
010	serial to RoboClaw controller

remaining codes available for future use

The next bit position (B3) specifies whether wheel encoders are available (1-yes, 0-no). The remaining bit positions may be used in the future.

139: SensorSetup

The three LSB's (B2,B1,B0) of the parameter provided specify the type of analog sensor used for the perimeter sensors in the virtual mode, or the ranging sensor in the digital mode as follows:

000	Sharp infrared short
001	Sharp infrared long
010	Maxbotics sonar
011	Parallax Ping)))

reserved for future types

The next three bits (B5,B4,B3) specify the compass used. Currently the HMC6352 is supported (code 001) and the HMC5883L (code 010) – 000 implies no compass.

The two MSBs (B7,B6) specify the mode as follows:

00	Digital perimeter sensors
01	5 ranging sensors, rear digital
10	6 ranging sensors
11	reserved for future use

140: SetReducForRight

The parameter provided sets a percentage to *reduce* the right wheel speed when going forward in order to correct drift. **Note**: These parameters should be set even when wheel encoders are used as closely matched wheels are easier to correct.

141: SetReducBackRight

The parameter provided sets a percentage to *reduce* the right wheel speed when going backward in order to correct drift.

142: SetReducForLeft

The parameter provided sets a percentage to *reduce* the left wheel speed when going forward in order to correct drift.

143: SetReducBackLeft

The parameter provided sets a percentage to *reduce* the left wheel speed when going backward in order to correct drift.

144: SetRROStimeout

The parameter provided controls the length of time the RROS will wait for a RobotBASIC command before shutting down the robot's drive motors. Normally this commands should be long enough that that the motors are not shut down between commands, but short enough that the motors will not run for any significant time if an application program ends without halting the robot. Normally a value of 15 or so is fine.

145: SetCountDivisor

The parameter specified is used to reduce the wheel encoder counts. For example, a divisor of three will decrease the actual count to $1/3$ of its actual count. This is only need if the counts generated by your hardware are too large for the 8-bit parameters specified for 90° turns and diameter-length moments.

146: SetDriveServWidth

The parameter specified can enable better linearity for servomotor powered robots by controlling the overall pulse width. Default value if 50. Range is 0-127.

147: FindBeacon

This command causes the robot to rotate until it finds a beacon or until the BeaconTimeOut has been exceeded. The parameter provided controls the direction of the rotation (0-left, 1-right).

148: SetSetBeaconDelay

The parameter provided determines the time the RROS will wait before starting its timing to determine which beacon is present. In general, this time period should wait until the middle of the first 500us beacon period. Note: This is provided only for those wishing to make their own beacons. The time is automatically calibrated for our beacon chips.

149: SetSteerServoWidth

The parameter provided determines the maximum pulse width for the steering servomotor in a steerable robot. The default value is 72.

150: SetDriveServoDir

The parameter provided reverses the direction of the drive servomotor in a steerable robot. Acceptable values are 0 and 1.

151: SetSteerServoDir

The parameter provided reverses the direction of the steering servomotor in a steerable robot. Acceptable values are 0 and1.

152: SetRobotDiameter

The parameter provided should be the diameter of the robot in ½ inch increments (a value of 10 means the robot's diameter is 5 inches). A non-zero value for this parameter causes the rRange readings to be similar in distance (relative to the robot's size) as that of the simulator.

153: SetTurretServoWidth

The parameter specified determines the pulsewidth for each degree of movement of the servomotor controlling a ranging turret. The default value is 18.

154: SetTurnToMax

The parameter determines the maximum speed (1-100) to use when executing the rCommand TurnToHalfAngle. **Note:** This speed is used when the destination is not close.

155: SetTurnToMin

The parameter determines the minimum speed (1-100) to use when executing the rCommand TurnToHalfAngle. **Note:** The speed of rotation will slow to this value as the desired destination is approached.

155

156: SetSlowDown2
The parameter determines the minimum speed used when counters are enabled. The robot's speed is automatically reduced to this speed when the robot is *very* close to its destination. The value should generally be as small as possible without stalling the robot.

157: SetPCBcalibrate
The parameter determines if the distance measured by Ping and SR04 sensors are adjusted for use with our PCB (which mounts the side-pointing sensors inward instead of on the edge of the robot's perimeter. Use values of 0 and 1 (or False and True) to disable or enable this feature (which defaults to enabled).

158: RROSversion
This commands has the RROS indicate the current version with a sequence of tones. Longer tones indicate the main number with shorter, quicker tones indicating the decimal. For example, two long tones followed by 3 short tones would indicate version 2.3. Current version is 2.1. Early versions of the RROS do not have this command.

159: Drive
The parameter for this command will drive the robot in one of four directions at a specified TurnStyle. The first 2 bits in the parameter control the direction of motion (The values for ForwRight, ForwLeft, RevRight, and RevLeft are defined in the file RROScommands.bas). The remaining 6 bits controls the percentage of the Speed that will be used for the slower wheel. Since there is only six bits, you must pass ½ the percentage (which limits the resolution to 2%). You could drive the robot forward right with the right wheel moving at only 50% of the left wheel with the following command.

rCommand(Drive, ForwRight+50/2)

160: ReadRangers
This command assumes you are using rangers to create the VSS (virtual sensory system). It returns the distance reading of the five main rangers (not the optional rear ranger) in a string. You can parse the string into an integer array as shown below. Note: This command is far faster than using rRange (five times) to read the five range values.

```
dim a[5]
x=rCommand(ReadRangers, 0)
for i=0 to 4
 a[i]=GetStrByte(x,i+1)
next
// the array a now contains the distances
```

161: SetCompassType
This command sets the compass type just as you would with SensorSetup. Especially useful when using the auto-setup for the RB-9 or Arlo (neither of which select a compass to avoid hanging if no compass exists).

162: ForwardWithSensorCheck
This command moves the robot forward at the current speed settings for a distance equal to the parameter passed. Normally all five bytes returned will be zero. If the robot detects an object with the rFeel sensors, it will slow down, stop moving, and return the actual distance traveled in the first byte of the five bytes returned. Note: Wheel encoders are required when using this command.

Codes 162-199 are reserved for main chip expansion

Expansion Chip Codes

200: ExpansionSetup

The RROS chip has been designed to interface with multiple external microcontrollers as described in the text. The parameter for this command establishes what types of controllers are currently available as described below. A bit value of 1 normally indicates the designated expansion is available. For the ARM expansion, two bits specify the type of expansion to use.

- B0   GPS expansion available
- B1   Camera (or equivalent) for rLook available
- B2   PEN control available
- B3   External sensors available
- B5,4   Arm expansion available (second RROS chip)
    - 01: Standard arm expansion
    - 10: Arm expansion with OUTPUT
    - 11: Arm expansion with NAV Assist
- B6   Custom expansion available

201: BecomeArmExpansion

This command is generally only used INTERNALLY but may be used in special cases (See Appendix C)

202: ReadExpSensors

If an external sensor process is available, the main RROS chip uses this command to obtain bump, proximity, and line sensor data from the external source and ORs that data with the standard sensor data. This command is generally only used internally. It is provided for those building a custom Sensor Expansion.

203: ReadArmSensors

If an arm expansion is available, this command is used INTERNALLY to reads the available (number available depends on arm expansion mode) rSense() bits and combines them with the one bit available on the main RROS chip. Appendix C discusses other ways this command can be used.

Codes 204-209 are reserved for RROS use.

Codes 210-255 are just passed to remote expansion chips with all five bytes obtained from an expansion returned to RobotBASIC.

Codes 210-229 are reserved for user custom expansions.

230: SetNAVbeacons

This command is used to establish the beacon number for each of the two LPS navigation beacons currently active. The upper nibble contains the number of the "green" beacon and the lower nibble contains the number of the "red" beacon as described in Chapter's 11 and 12.

231: ReadNAVangles

This command requests the external NAV assist read and report information about the angles to the "green" and "red" beacons. The servomotor's pulsewidth for the green angle is reported high-byte first in the first two bytes of the five bytes returned. The pulsewidth for the red beacon is in the next two bytes. See Chapter's 11 and 12 for converting this information into angles and the robot's x,y position in the room.

232: ReadArmAnalogs

This command returns the data from five analog ports on the arm expansion chip as described in the text.

233: SetServoNumber

The external arm controller (a second RROS chip) can control five servomotors using this command's parameter as the index for the servomotor commands that follow. Note: The servomotor commands acting on an indexed position will automatically increment the index after they perform their function.

234: SetServoPosition

The parameter provided in this command establishes a new position for the currently indexed servomotor.

235: SetServoMin

The parameter provided in this command establishes the minimum pulse size applied to the servomotors for Position 0. The parameter can vary from 0 to 255.

236: SetServoSpeed

The parameter provided in this command establishes a new speed for the currently indexed servomotor.

237:   SetServoMax

The parameter provided in this command establishes the maximum pulse size applied to the servomotors for Position 255. The parameter should generally vary from 0 to 100 (%) but a small amount of over pulsing is allowed when needed.

238: EnableServos

This command enables (or disables) the arm servomotors with a true/false parameter. The default mode is disabled to ensure the arm does not move until commanded to do so. When the arm is enabled, the control pulses for all motors will be slowly increased in frequency to slowly move the arm to its initial position, regardless of where the arm resides at startup. Even with this attempt at initializing the servomotors slowly, it is recommended that the user always leave the arm in a known position and initialize the starting positions accordingly.

239: GPSsetup

The parameter provided is sent to a remote GPS controller if available. The parameter provided can be used in any way the controller wished. Example: The parameter could specify the units to use when specifying a position (feet, meters, etc).

240: SetGPShighX =   239

The parameter provided is sent to the GPS controller to be used as the initial value for the upper byte of a 16-bit data word representing the starting x position of the GPS system.

241: SetGPSlowX

The parameter provided is sent to the GPS controller to be used as the initial value for the lower byte of a 16-bit data word representing the starting x position of the GPS system. This command automatically clears the high byte, so it should always be used first when setting the starting x coordinate (meaning you do not have to specify the upper byte if it is zero).

242: SetGPShighY

The parameter provided is sent to the GPS controller to be used as the initial value for the upper byte of a 16-bit data word representing the starting y position of the GPS system.

243: SetGPSlowY

The parameter provided is sent to the GPS controller to be used as the initial value for the lower byte of a 16-bit data word representing the starting y position of the GPS system. This command automatically clears the high byte, so it should always be used first when setting the starting y coordinate (meaning you do not have to specify the upper byte if it is zero).

244: ExtSensorSetup
The parameter provided is sent to the external sensor controller (if one exists) to provide whatever initialization information is required.

245: LookSetup
The parameter provided is sent to the external rLook() controller (if one exists) to provide whatever initialization information is required.

246: PenSetup
The parameter provided is sent to the external rPen() controller (if one exists) to provide whatever initialization information is required.

247: SetArmOutputs
The expansion arm controller can have from 1 to 3 output bits (based on the mode established by BecomeArmController). The parameter provided will provide new values for the available bits as discussed in the text.

248:SetNAVservoMinHigh
The parameter provided is sent to the arm expansion chip to set the high-byte value of the minimum pulse width used by the navigation assist servomotor. See text for details.

249:SetNAVservoMinLow
The parameter provided is sent to the arm expansion chip to set the low-byte value of the minimum pulse width used by the navigation assist servomotor. This command clears the high-byte so it should be used first (allowing the high byte not to be written if zero).

250:SetNAVservoMaxHigh
The parameter provided is sent to the arm expansion chip to set the high-byte value of the maximum pulse width used by the navigation assist servomotor.

251:SetNAVservoMaxLow
The parameter provided is sent to the arm expansion chip to set the low-byte value of the maximum pulse width used by the navigation assist servomotor. This command clears the high-byte so it should be used first (allowing the high byte not to be written if zero).

Codes 252-255 are reserved for RROS expansions.

Main RobotBASIC Commands
RobotBASIC automatically sends most simulator command to the RROS after an **rCommPort** command has been issued, as described in the RobotBASIC HELP file. All commands have a two-byte format with five bytes being returned. For example, the command **rForward 30** will send the code 6 followed by 30 out over the port specified with the **rCommPort** command. Knowing the codes used for these commands can be valuable for non-RobotBASIC users. For example, your main programs could be run on an Arduino processor that simply uses the RROS chip as an intelligent peripheral. Appendix C will discuss this idea further. Figure A-1 provides a summary of the RobotBASIC command protocol.

Command	Op-code	Returned Bytes
rLocate	3	bump,prox,line,0,0
rForward	6	bump,prox,line,0,0
(backward)	7	bump,prox,line,0,0
rTurn (right)	12	bump,prox,line,0,0
(left)	13	bump,prox,line,0,0
rCompass	24	bump,prox,line,degrees
rBeacon	96	bump,prox,line,T/F
rLook (right)	48	bump,prox,line,color
(left)	49	bump,prox,line,color
rPen	129	bump,prox,line,0,0
rRange (right)	192	bump,prox,line,dist.
(left)	193	bump,prox,line,dist.
rSpeed	36	bump,prox,line,0,0
rChargeLevel	108	bump,prox,line,0,0
rGPS	66	x,y,0

Distance, degrees, color, x, and y are 2-byte integers (MSB first)

Figure A-1: These examples of RobotBASIC commands (and their op-codes) are implemented in the internal wireless protocol.

Motor and Sensor Setup Parameters (& misc.)

0: SMALLDC
1: SERVOMOTORS
2: ROBOCLAW
3: STEERABLE
0: IRSHORT
1: IRLONG
2: MAXBOTICS
3: PING
4: SR04
8: HMC6352
16: HMC5883
8: ENCODERS
0: DIGITAL
64: FIVERANGE
128: SIXRANGE
1: GPS
2: LOOK
4: PEN
8: EXTSENSORS
16: ARMEXPANSION
1: ARM
2: ARMwOUT
3: ARMwNAV
0: NORTH
180: SOUTH
90: EAST
270: WEST
253: ARLO // optionally used as 1st param

160

254: RB9 // in rLocate for faster initialization
255: SlowComm
0: ForwRight
64: ForwLeft
128: RevRight
192: RevLeft

Sound Parameters

0: Pause
1: LowC
2: LowD
3: LowE
4: LowF
5: LowG
6: LowA
7: LowB
8: MidC
9: MidD
10: MidE
11: MidF
12: MidG
13: MidA
14: MidB
15: HighC
16: HighD
17: HighE
18: HighF
19: HighG
20: HighA
21: HighB
22: Blip1
23: Blip2
24: InitTone
25: LowTone
26: BeepBeep
27: BeepBeepBeep
28: Phasor
29: Siren1
30: Siren2
31: Siren3
32: Quarter
64: Half
128: Whole
224: Double

```
// Partial song for illustration purposes
// To play, mcopy to CurrentSong and call PlayMySong
Data Birthday; MidC|Quarter,MidC|Quarter,MidD|Half,MidC+Whole
Data Birthday; MidF+Half,MidE+Half,Pause+Half,MidC+Quarter Data Birthday; MidC+Quarter,MidD+Half,MidC+Whole,MidG+Half
Data Birthday; MidF+Half,Pause+Half MidC+Quarter,MidC+Quarter Data Birthday; HighC+Half,MidA+Half,MidF+Half,MidE+Half
Data Birthday; MidD+Whole,Pause+HalfMidB+Quarter
Data Birthday; MidB+Quarter,MidA+HalfMidF+Half,MidG+Half
Data Birthday; MidF+Whole,0

// subroutine to play CurrentSong
PlayMySong:
 NotePosition=0
 while(1)
  if CurrentSong[NotePosition]=0 then break;
  rCommand(PlaySound,CurrentSong[NotePosition])
  NotePosition++
 wend
return
```

Quick Reference
Hardware Configuration Summary

This Appendix provides a quick description of how each pin on the RROS chip is used in both the standard mode and arm expansion mode. **REMEMBER**: This is just a summary. Refer to the text for DETAILS.

Standard RROS Pin-outs

RROS Pins 1,2:
These two pins are connected to the left drive motor as follows:
- to the DC drive motor's leads when the onboard motor controller is used
- to the control input on a servomotor when used (**either** pin may be used for this purpose – DO NOT short them together), pull-up resistor required

RROS Pin 3:
This pin is the input for the left wheel encoder.

RROS Pin 4:
This pin is the input for the right wheel encoder.

RROS Pin 5:
This pin is the input for the rSense() bit B0.

RROS Pin 6:
This pin has two uses:
- Normally, it is the input for the rSense() bit B1.
- When an expansion is used, it is the serial receive pin.

RROS Pin 7:
This pin has two uses:
- Normally, it is the input for the rSense() bit B2.
- When an expansion is used, it is the serial transmit pin.

RROS Pin 8:
This pin is the receive pin for communicating with RobotBASIC (usually Bluetooth or other wireless).

RROS Pin 9:
This pin is the transmit pin for communicating with RobotBASIC (usually Bluetooth or other wireless).

RROS Pin 10:
This pin has many uses. All of the following possibilities are mutually exclusive. It connects to:
- the MUX relay when six analog perimeter sensors are used
- to a rear digital sensor when only five ranging sensors are used
- to a rear Ping sensor (Position P6) when six Pings are used
- to the turret's servomotor in the all-digital mode

RROS Pin 11:
This pin connects to both the beacon detector and the piezo buzzer.

RROS Pin 12:
This pin has many uses. All of the following possibilities are mutually exclusive. It connects to:
- the trigger when Maxbotics ultrasonic sensors are used
- to the Ping sensor at position P5 when Pings are used
- to the digital sensor position at D5 if a turret mounted Ping is not used in the all-digital mode
- to the Ping ranger when used in the all-digital mode

RROS Pin 13:
Hard reset, not used.

RROS Pin 14:
This pin has many uses. All of the following possibilities are mutually exclusive. It connects to:
- the analog ranging sensor at position A5 in the non digital mode
- the turret mounted analog ranging sensor in the all-digital mode
- the digital sensor at position D5 if a turret mounted Ping is used

RROS Pin 15:
This pin connects to the perimeter sensor at position 1 whether it is analog, digital or Ping.

RROS Pin 16:
This pin connects to the perimeter sensor at position 2 whether it is analog, digital or Ping.

RROS Pin 17:
This pin connects to the perimeter sensor at position 3 whether it is analog, digital or Ping.

RROS Pin 18:
This pin connects to the perimeter sensor at position 4 whether it is analog, digital or Ping.

RROS Pin 19:
This pin connects to the data line of an I^2C supported compass.

RROS Pin 20:
This pin connects to the clock line of an I^{2}C supported compass.

RROS Pins 21,22:
These two pins have several functions:
- to the right DC drive motor's leads when the onboard motor controller is used
- to the control input on the right servomotor when used (**either** pin may be used for this purpose – DO NOT short them together), pull-up resistor required
- to the serial input on a RoboClaw when used to power large DC motors (**either** pin may be used for this purpose – DO NOT short them together), pull-up resistor required

RROS Pin 23:
Main system ground.

RROS Pin 24:
Main system power (6-12 volts). Also supplies power to small DC motors though the onboard controller when used.

RROS Pin 25:
This pin can connect to:
- the main battery voltage passively divided to approximately 4 volts (when six analog sensors are not used)
- to relay contacts that multiplex a sixth analog sensor with the battery voltage as described above

Arm Controller Expansion Pin-outs

The RROS chip can be instructed to become an arm controller chip. In that mode, the pin-outs are as follows. Remember, this is just a summary. Refer to the text for details.

Arm Expansion Pins 1,2:
Not used

Arm Expansion Pin 3:
Connects to a servomotor control line.
Recommended: Robot arm shoulder joint.

Arm Expansion Pin 4:
Connects to a servomotor control line.
Recommended: Robot arm elbow joint.

Arm Expansion Pin 5:
Connects to a servomotor control line.
Recommended: Robot hand open/close.

Arm Expansion Pin 6:
Connects to a servomotor control line.
Recommended: Robot arm wrist up/down.

Arm Expansion Pin 7:
Connects to a servomotor control line.
Recommended: Robot arm wrist rotate .

Arm Expansion Pin 8:
Serial receive pin for communication with the main RROS chip.

Arm Expansion Pin 9:
This pin has two uses as follows:
- it is the least significant output bit when the NAV assist mode is not selected
- it connects to the NAV assist servomotor control line when that mode is selected

Arm Expansion Pin 10:
This pin serves as the input for rSense() bit B1.

Arm Expansion Pin 11:
This pin serves as the input for rSense() bit B2.

Arm Expansion Pin 12:
This pin serves as the input for rSense() bit B3.

Arm Expansion Pin 13:
Analog input 4 of 5.

Arm Expansion Pin 14:
System reset. Not used.

Arm Expansion Pin 15:
This pin serves as the input for rSense() bit B4.

Arm Expansion Pin 16:
This pin has three uses as follows:
- in the normal mode it is rSense() bit B5
- in the Output mode it is output bit B1
- in the NAV assist mode it is the input for the wide beacon sensor

Arm Expansion Pin 17:
This pin has three uses as follows:
- in the normal mode it is rSense() bit B6
- in the Output mode it is output bit B2
- in the NAV assist mode it is the input for the narrow beacon sensor

Arm Expansion Pin 18:
Analog input 1 of 5

Arm Expansion Pin 19:
Analog input 2 of 5

Arm Expansion Pin 20:
Analog input 3 of 5

Arm Expansion Pins 21,22:
These pins can serve as the serial transmit output to communicate with the main RROS chip. (**either** pin may be used for this purpose – DO NOT short them together), pull-up resistor required.

Arm Expansion Pin 23:
Main system ground.

Arm Expansion Pin 24:
Main system power (6-12 volts)

Arm Expansion Pin 25:
Analog input 5 of 5.

Using the RROS with
Arduino and Other Microcontrollers

Although the RROS was designed specifically to interface with RobotBASIC, it can be used as a general purpose hardware interface for almost any microcontroller or system including the BASIC Stamp, Picaxe, Baby Orangutan, and Arduino processors.

Basically, the controller to be used simply needs to send out a 2-byte command (serially at 9600 baud) and wait for a 5-byte reply. The details of both of these communications are summarized in Figure A-1 of Appendix A.

It is assumed that users of controllers such as those listed above are familiar with their programming languages and the communication routines available through their libraries. Because of this, and because space does not permit addressing specific examples for every possible processor, we will provide a simple RobotBASIC example to demonstrate the techniques involved. We hope that after seeing how easy the process is, that users of other processors can proceed on their own. In fact, we invite users to submit their programming endeavors along these lines and we will make some of them available on our webpage or perhaps even small examples in this manual.

Our sample RobotBASIC programs will assume that we have one communication subroutine, CommandOut(), that can transmit our commands and wait for sensory data via SensoryDataIn(). The exact structures could be very different from our demo code. For example, if you are using C to program your processor, the 2-byte commands and 5 bytes of returned data could each be stored in arrays. It is also important to realize that the library functions provided by many controllers already provide much of the desired functionality implemented by our code. Figure C-1 shows how these subroutines can be built with RobotBASIC.

```
sub CommandOut(Cmd, Param)
  // send command to the RROS chip
  SerOut(char(Cmd))
  SerOut(char(Param))
  // then wait for returned data
  call SensoryDataIn()
return

sub SensoryDataIn()
```

```
// It is assumed that a 5-element array
// CommData[] already exists.
repeat  // wait until 5 bytes are received
  SerBytesIn 5,CommData,n
until n=5
return
```

Figure C-1: These two subroutines handle all communications with the RROS chip.

In order to demonstrate how to utilize the communication routines lets assume we want to initialize the robot, move it forward 100 units, turn it 90° to the right, and move forward until some object is detected in *front* of the robot by any of the perimeter sensors (rBumper or rFeel). We will also assume that the robot is powered by small DC motors and that it uses a Virtual Sensor System composed of six Ping sensors. Sample code to perform these operations is shown in Figure C-2.

```
Main:
 gosub Initialize
 call CommandOut(6,100) // forward 100
 call CommandOut(12,90) // right 90 degrees
 repeat
  call CommandOut(6,1) // forward 1
  call FormatData(Bump,Feel,junk,junk,junk)
  // variable junk used for unneeded data
 until (Bump&4) OR (Feel&14)
end


Initialize:
 Dim CommData[5] // array for sensory data
 SetCommPort 47 // use number for your port
 call CommandOut(3,0) // rLocate
 call CommandOut(138,0) // DC motor setup
 call CommandOut(139,131) // Ping sensor setup
 // add any fine tuning here
return


sub FormatData(&a, &b, &c, &d)
 // re-format array into the parameter list
 for j=1 to 4
  y = ascii(substring(CommData,j,1))
  if j=1
   a=y
  elseif j=2
   b=y
  elseif j=3
   c=y
  elseif j=4
   d=y*256+ ascii(substring(CommData,5,1))
  endif
 next
return
```

Figure C-2: This code moves a robot as indicated in the text.

Study the codes used by the RROS to fully understand this program. At that point you should be able to utilize the RROS chip with nearly any microcontroller.

Notice how the function FormatData() is used to extract data from the sensory array. The first four variables in the parameter list become byte versions of the first four elements in the sensory array. The last becomes the integer version of the last two bytes in the array to handle things like compass readings. For example, the following code fragment will read the compass heading from the RROS chip along with the rBumper(), rFeel(), and rLine() data that is also available when the compass is interrogated.

```
call CommandOut(24,0) // reads compass data
call FormatData(Bump,Feel,LineData,Heading)
```

An Arduino Example

In order to help clarify the process of converting our RobotBASIC example code (think of it as pseudo code) to other languages, let's look at Figure C-3.

```
char CommData[5];

void SensoryDataIn(void)
  {
  // wait till 5 bytes have arrived
  while(Serial.available()<5)
    ;
  for(int j=0;j<5;j++)
    {
    CommData[j]=Serial.read();
    }
  }

void CommandOut(char c, char p)
  {
  Serial.write(c);
  Serial.write(p);
  SensoryDataIn();
  }

void Initialize(void)
  {
  Serial.begin(9600); // init arduino serial
  CommandOut(3,0);   // rLocate
  CommandOut(138,0); // motor setup
  Serial.write(139,131);  // sensor setup
  }
```

Figure C-3: An Arduino version of the main routines.

Figure C-3 shows a straightforward implementation of the low-level functions used in the RobotBASIC example, but converted to work on most Arduino controllers. As you can see, the translation is relatively painless as long as you understand the operations required, and have a working knowledge of the target language.

Arm Expansion

You can even use the RROS chip in the Arm-Expansion mode with other processors. Communicating with the chip in this mode is identical as the main RROS communication (always 2 bytes to the chip, always 5 bytes returned from the chip).

You can send all Arm related codes directly to the chip after it has been told to convert itself to an arm expansion chip. You can make the RROS chip alter its identity and become an Arm Expansion chip by sending it the code 201 (normally ONLY used by RROS internally) followed by a 0,1,or 2 as dictated as below.

> 0 – Normal arm mode
> 1 – With 3 output bits
> 2 – With NAV assist

Another code that is used internally by the RROS is 203, which reads the available rSense() sensors (digital). The data will be returned in the third byte of the five bytes returned.

Assembling the RB-PCB

Although the RROS chip can easily be used to build a robot using solderless breadboards, the RB-PCB (RobotBASIC-Printed Circuit Board) makes the job even easier. Robot's built this way are also less likely to have problems when handled or transported.

The PCB may be updated from time to time, but the photo's in this Appendix should be substantially the same as the one you are using. If you have questions or concerns about your board please email us.

This the first portion of this Appendix assumes you are assembling the board using our Parts Kit. The board DOES support other sensor and motor configurations and they are discussed later in this Appendix.

Usually it is important to use sockets for most of the larger components. This allows them to be replaced when damaged but it also allows you to use other components when the need arises. Let's look at one example.

The Voltage Regulator
We suggest using a 7805 regulator, and that is the one provided in our kit of parts. We also suggest using a 6-volt gel-cell battery because it has excellent power for the price and it does not require an intelligent charger like many other batteries.

The problem is that the 7805 requires more than 6 volts in order to ensure that its output is a true 5V. Fortunately though, our tests have shown that the lower regulated voltage you get from a 7805 when used with a 6-volt gel-cell (which actually produces about 6.6 volts) as its input does not seem to cause any problems with any of our prototypes. If you substitute components, and were to have problems because of the slightly lower supply voltage, and you used a socket with the 7805, then you could replace it with a LM2940 regulator, which has the same pinout.

Motor Configurations
The RROS chips can drive small DC motors directly without any additional components. We utilize this capability with our RB-9 chassis. The motors are small enough to be driven by the chip. They also contain integrated wheel encoders that provide pulses that the RROS counts as wheels turn, allowing the movements to be far more accurate that open-loop timed responses.

Small DC Motors
The solder pads at the bottom right side of the RB-PCB provide connections for small DC motors. In our parts kit we provided four single pin male headers that allow the motor cables from the RB-9 chassis to plug directly into these positions (the RED lead of both motors should be on the left).

If you are soldering your motors, remember that the leads may have to be reversed if your motors run backward.

Servomotors

Since the RROS also supports continuous rotation servomotors, the RB-PCB has solder pads for two 3-pin male headers (or one 6-pin) on the left side of the board, above the line sensor headers. You can plug standard sized servomotors directly into these two headers. Larger servo motors might require a special cable that allows the motors to be powered from a separate power supply. We have limited options for this built into the PCB, so contact us for more information if this situation applies to you.

There are four 10K resistors on the far left side of the RB-PCB. The top two of these act as pull-ups for the I2C compass. The lower two resistors provide the necessary pull-ups when the RROS drives servomotors. While these resistors should not interfere with operations when other motors are used, it is recommended that they not be installed unless needed.

The Roboclaw Motor Controller

The RROS can utilize a Roboclaw motor controller to handle motors up to 30 amps. The right servomotor connector (see above) can be used as the serial connection to the Roboclaw (see Chapter 5 for more details). The 10K pull-up resistors required for servomotors should also be installed when a Roboclaw is used.

The Sockets

The first step is to insert inline female sockets into all the appropriate positions. Figure D-1 shows the sockets we provide with our parts kit for the RB-PCB. The photo also shows a number of inline male headers. **Note:** We provide five 4-pin sockets in the parts kit for perimeter sensors. If you wish to use digital or Ping sensors you can request that we substitute 3-pin sockets or pins for these five parts.

The Arlo robot mentioned throughout this manual often needs cables between the RROS board and the motors and various sensors. Because of that, it is often better to use male pins instead of female sockets. If there is enough interest, we may offer an "Arlo Parts Kit" with more appropriate connectors. For more specific information about hardware connections for Arlo, refer to out book *Arlo: The Robot You've Always Wanted* (available on Amazon.com Summer 2015).

Figure D-1 also shows a number of inline male headers. Note the four single pin male header pins used for the motor connections. These allow the motors from the Rover 5 and our RB-9 chassis to connect directly to the board. You may need to solder or provide alternative connectivity based on your needs.

When installing the sockets, here is a suggestion. Hold the board in your hand while inserting all of the headers. When finished, cover the headers with a piece of cardboard and turn the board over and lay it on a flat surface. This will keep all the headers in place while you solder them. In order to make this job even easier, we have specifically tried to keep the hole sizes for the headers as small as possible because that helps you keep everything together while soldering (even though this makes the holes too small for some components).

Figure D-1: Start your assembly by inserting the male and female headers.

Power Connection

Figure D-2 shows the screw terminal block for connecting the PCB to the battery. It also shows an on/off switch for the board. In many cases this switch is all you need. In some cases though, the PCB might be difficult to access. In such cases, just leave this switch in the ON state and add an external switch in series with the battery wiring. We have provided several holes in our RB-9 chassis for such a switch.

Figure D-2: Screw terminals provide easy connection to the battery. An on/off switch is also integrated into the board.

Resistors

The next step is to solder in the resistors as shown in Figure D-3. The two lower 10K resistors on the left side of the board need only be installed if you are using servomotors or a Roboclaw as discussed earlier in this chapter. We supply 10K resistors for R1 and R2 but you can alter them if you wish – refer to Chapter 9 for more information.

Capacitors

The 4.7uf capacitor just to the left of the beacon detector socket (see Figure D-3) is the only capacitor that is actually necessary based on all our tests. We know though, that you may use our board with other motors that might have significant electrical noise, at least compared to our recommended motors. You might also be using your robot in an environment where electrical noise is present. For these reasons, we have added mounting holes for six optional capacitors. We do not think these will ever be necessary, but it did not cost anything to provide this option for you in case the need ever does arise.

Even though our tests showed that none of these optional capacitors are needed, we have included three (#'s 2, 3, and 6) in the parts kit to help prevent any problems we might have overlooked. Figure D-3 shows these capacitors in place. The actual values are not critical, especially since no specific need for these capacitors has been established. Just use the ones supplied in the kit. Typically Cap 2 is a .1uf ceramic disk and Caps 3 and 6 are 100uf electrolytics. If you ever have noise problems, contact us and we will make recommendations for the other capacitors based on your situation.

Figure D-3: This completes the soldered components.

Wheel Encoders

Wheel encoders can provide pulses indicating how far your robot's wheels have moved. If you have them, they can be connect by cable to the male headers labeled L.CNT and R.CNT (for the left motor counter and the right motor counter). The R (red) and B (black) labels indicate +5V and ground that can be used to power your encoders. Connect your encoder data output for each motor to the appropriate D pin.

Most wheel encoders are quadrature encoders, which have two sets of ouputs for each wheel. The RROS only needs one set of pulses for each wheel – either quadrature output can be use. Just leave the other disconnected.

Remaining Components

Finally, solder in the transistor, the buzzer and the transistor as shown in Figure D-3. Figure D-4 shows a completed board with the RROS chip, the compass, the Bluetooth transceiver, the beacon-detector, and the five sonic sensors in place. There are several things that need further discussion.

Figure D-4: The finished board provides more sensors than most hobby or educational robots.

Buzzer Mounting

There are multiple mounting holes for the ground connection of the buzzer (sound transducer). One of these configurations should fit many common parts including the one we ship. If you use an odd-sized transducer you may have to solder extra wires to it or bend its leads to make it fit the provided holes.

The Transistor

We supply a 2N697 NPN transistor in our kit. The transistor provides buffering so that RROS Pin 11 can be used to drive the buzzer AND read data from the Beacon Detector. The specifications for the transistor are not critical and nearly any general purpose NPN transistor should work fine.

The Compass

The mounting hole configuration for the compass uses both a 3-pin and a 4-pin socket. This allows you to use the 6-pin Parallax style 6352-based compass or the 4-pin version of the 6352 currently provided by most vendors. Refer to the lettering on the RB-PCB to see how the compass should be oriented.

The 6352 compass has been around for many years. It works great, but there is always a chance that older parts may be discontinued. For that reason, the RROS also supports the HMC5883L compass (available from Parallax, RobotBASIC, and others). Typically the 5883 has 5 interfacing pins – only 4 are needed (+V, Ground, Clock, and Data). There are several pin configurations available for the 5883, so you need to use a cable with individual connectors to connect it to the RROS board.

If IR sensors are used, or if the perimeter sensors are cabled away from the PCB, then a socketed compass might work fine mounted on the PCB. When ultrasonic sensors are used though, the compass itself MUST be mounted above the sonic sensors so that electrical noise and magnetic fields will not interfere with the compass readings. The best way to do this is to solder wires to the extra provided sockets as shown in Figure D-5. Inserting these wires into the sockets on the board, effectively extends the sockets upward raising the compass to the desired level (see Figure D-6). You need two extended sockets for mounting the Parallax style compass but only one (the 4 pin) for the other versions of the compass. It is important that the compass be LEVEL in order to get accurate readings.

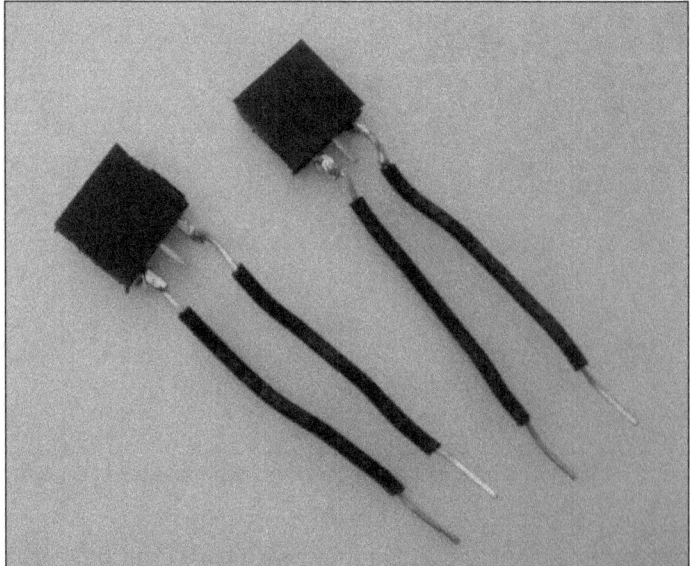

Figure D-5: If you prepare these two extenders, you can use them to mount either of the supported compass configurations.
NOTE: In some cases, one of these extensions should be a 4-pin socket with 4 wires for the inline compass.

Of course, you do not have this problem with the 5883 compass because it has to be cabled away from the PCB anyway. Do make sure that you mount it in a level position, as the horizontal position dramatically affects the readings. You can mount in any direction though since you can use rCommand, SetRobotAngle to add an offset to the actual compass reading.

Figure D-6: Notice how the compass socket is extended.

Rear Sensor

You may use a digital rear sensor (such as the Pololu Item #1134) or even a physical bumper switch with the SR04 configuration. The rear sensor should connect to the REAR 1 connector. We offer cables that connect directly to the Pololu Sensor mentioned above.

Line Sensors

Three line sensors are supported in all of the single RROS Chip modes. You can use any IR reflective sensors as long as they have a digital output such as the Pololu Item # 958 (NOTE: Pololu calls this an *analog* sensor – their *digital* sensor will not work with the RROS chip.) We offer cables for connecting these sensors to the line-sensor sockets.

If you do not need line sensors, you can utilize these inputs for any digital sensory input your robot might need.

Pin 25

The wire coming from Pin-25 on the RROS chip should plug into the battery monitoring socket near R1 and R2.

179

The Beacon Detector

A 3-pin male header is provided in the kit of parts. You should solder the Beacon Detector to this header so that it can be inserted in the socket for it (see Figure D-7). This allows you to remove the header to better create a hood for it. It also ensures that the detector is high enough to see over the front sonic sensor. **Note:** The size of your hood depends on the size of your robot's environment etc. In general, a 3/16 inch slot is reasonable as discussed in Chapter 12.

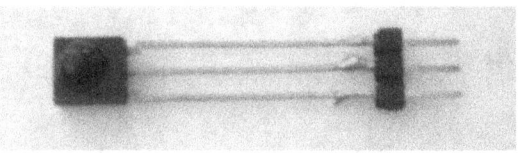

Figure D-7: A 3-pin male header is soldered to the beacon detector.

The Trigger Jumper

There is a Trigger-Jumper-Block just above the DC motor connectors. When using the SR04 Sonic Sensors place a shorting block (provided in our kit) across the two pins on the end labeled SS.

Alternate Sensor Configurations

If you have read the entire RROS manual you know the RROS chip allows you to build a robot with six different types of perimeter sensors our PCB supports nearly all the possible modes. Remember, you MUST always initialize the RROS chip with the appropriate Sensor Setup Code as described in Chapter 9.

It is IMPORTANT that you realize that connecting sensors in the optional modes may get a little confusing, depending on the mode. There are just so many options for all these modes that the connections are not always straightforward. Even so, we feel our RB-PCB provides acceptable options for all of the supported sensory modes.

Ping Sensors

If you wish to use Ping sensors, their 3-pin sockets are directly in front of the 4-pin sockets used for the SR04 sensors. No solder holes are provided for Ping Sensors, so sockets (or male pins) should be used. Generally we can substitute 3-Pin Ping sockets for 4-pin SR04 sockets if you ask us to do so when a parts kit is ordered. We can even supply both sets for a small additional charge.

In this mode either a digital rear sensor (such as the Pololu Item #1134) or a rear Ping sensor can be connected to the Rear 2 connector using a cable.

Maxbotics Sensors

The pin configuration for the Maxbotic Sensors is much different from the other supported sensors. We assume that most people will only choose the Maxbotics option if they are building a larger man-sized robot. For that reason, each Maxbotics Sensor must be connected to the board using 4 male-female connecting cables that allow you to mount the sensors away from the RB-PCB at positions appropriate for your robot.

Two connecting wires should provide power and ground from the appropriate SR04 socket pins. The Maxbotic Trigger pin (RX) should be connected to the trigger pin on the SR04 socket. You MUST move the Trigger-Jumper to the **Max** end of the trigger jumper block.

The Maxbotics Data pin is labeled AN and should be connected to the D pin on the SR04 socket for all sensors EXCEPT the left-pointing sensor. In that case, the D pin should connect to the RB-PCB pin marked Analog D (for analog data). A digital rear sensor like the Pololu Item #1134 can be used in this mode by connecting it to the REAR 2 connector with a cable we offer.

If you wish to use a Maxbotic sensor on the rear, you can use the REAR 2 signal to control a separate, user provided, relay that switches the RROS Pin 25 input between the battery monitoring signal and the output from the rear sensor (see Chapter 9). The signal from Rear 2 should be used to activate the relay.

Pololu IR Rangers
The supported Pololu IR rangers have connecting wires instead of pins. For that reason, they should be soldered to the pads directly behind the SR04 sockets in the following manner.

The RED and BLACK wires should connect to the + and G terminals (respectively) for each socket. The white wire from each sensor should connect to the D terminal on every socket except for the left-pointing sensor, where it should be connected to the Analog D pin as we did with the Maxbotic sensors.

The rear sensor can be either a digital sensor or another IR ranger.

Digital Mode
If you wish to use 5 digital sensors (instead of implementing a full VSS as described in Chapter 9) then the IR sensors (Pololu Item #1134) can be connected to the sockets for the Ping sensors directly (or with cables if you prefer).

This mode also supports a ranging sensor mounted on a servomotor controlled turret. The servomotor can be connected to Rear 2 connector with any of the supported analog rangers cabled to the Rear 1 connector (the Maxbotics must be wired as no cable is available).

If you wish to use a Ping Ranger for this mode, then it must be connected to the Ping socket for the left-pointing perimeter sensor and the left-pointing perimeter sensor must be cabled to Rear 1. (This is such a convoluted option we generally don't recommend it.)

Problems with Echos
Chapter 9 discusses the idea that echos associated with one ultrasonic sensor can be detected by other sensors – especially when the robot is small and the sensors are close together. There are several solutions for this problem.

First, IR sensors can be used on small robots instead of ultrasonic ones. If ultrasonic is desired, then Ping sensors may be preferred over SR04's even though the Pings cost considerably more. Pings have the ability to be triggered through their data line which allows the RROS to read the sensors in two groups that do not interfere with each other.

With other ultrasonic sensors, you can mount them near the edges of the robot using cables. If the robot is large enough, they can be placed so that they do not interfere with each other. If you want to keep the SR04 sensors on the board (as shown in Figure D-4) you must make them more directional using a hood to shield them from unwanted echos. Figure D-8 shows hoods we made for the SR04 sensors.

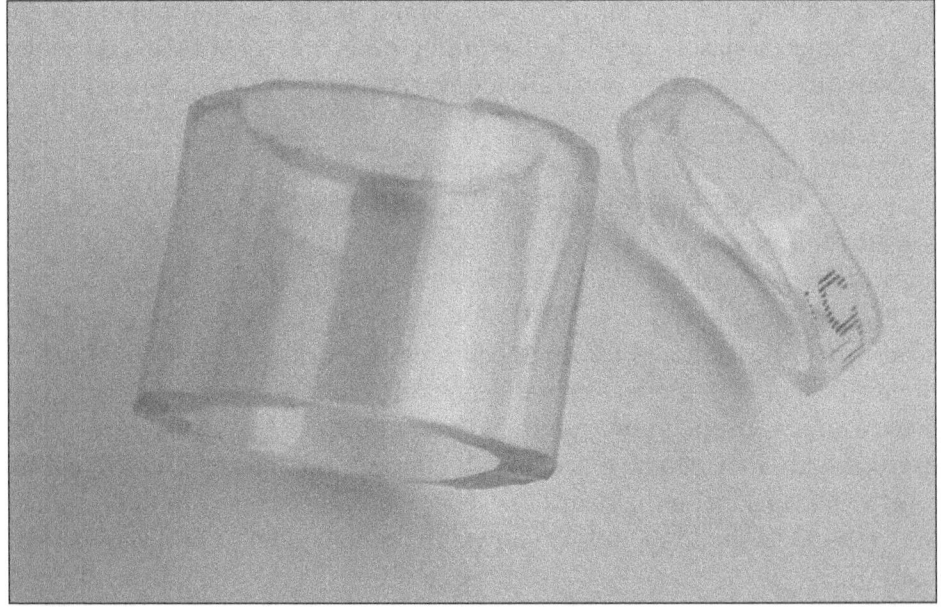

Figure D-8: False reading can be minimized by placing hoods on the receiving transducer as shown above.

The hoods are made from rubber tubing found at most hardware stores as shown in Figure D-9. The larger tube is 1 inch OD and the smaller .75 inch OD. The smaller tube fits snuggly inside the larger one.

Figure D-9: Two small pieces of rubber tubing can be used to make a hooded shield.

The smaller tube also fits snuggly over the SR04 sensor as shown in Figure D-10. Make sure the small tube does NOT protrude past the front edge of the sensor.

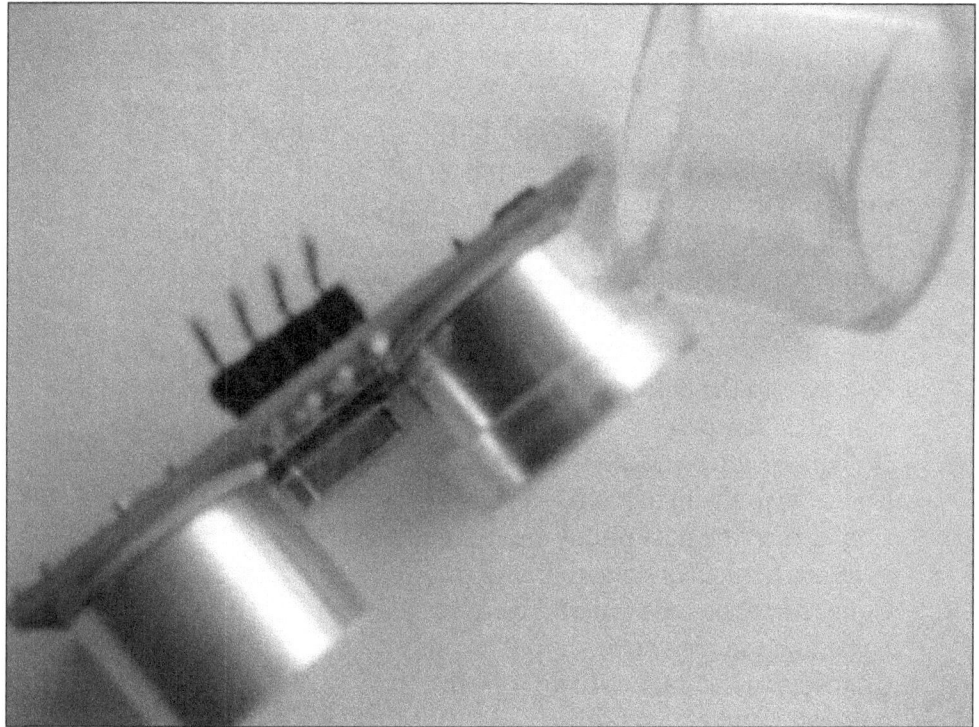

Figure D-10: The small tube fits over the sensor, then the larger tube fits over it.

The purpose of the smaller tube is to ensure that the larger tube is well away from the pathway of the echo signals. The larger tube can be pressed over the smaller tube as seen in Figure D-8 (take care not to twist too much and damage the sensor). The larger tube should extend about .5 inch out from the front edge of the sensor. Too much or too little extension will not work properly – just experiment till you get it right (use a small program that constantly displays the distance measure by each sensor). It is also important that you use the soft tubing as shown in our figures. Harder materials can reflect the sound waves and render the hood ineffective. **Note:** Often the tubing in the store has been squeezed on the roll making it far from round. After cutting the tubes, place them in VERY hot water for a short time to help get them pliable and round again. They must be round in order to work without interfering with the operation of the ultrasonic sensors.

No Schematic
For the most part, a schematic of the PCB would provide little value. This is true because the board generally does not provide any additional circuitry. Instead it simply provides connectivity to make it easy to utilize most of the supported sensors. The RROS pin connections below should provide the information most people need.

RROS PINS	Connection
1,2	Connects to the Left Motor (and one connects to the left servomotor header)
3,4	Connect to the left and right wheel encoder headers on each side of the RROS chip
5,6,7	Connect to the Line sensor headers (Pin 7 on left)
8	Connects to the Bluetooth Xmit pin
9	Connects to the Bluetooth Rec pin
10	Connects to the center of the Trig-jumper block, and to the data pin on Rear 2
11	Connects to the data pin of the Beacon Detector and to the transistor buffer that drives the sound transducer.
12	Connects to the data pin of the left pointing perimeter sensor.
13	No connection
14	Connects to the data pin of Rear 1
15	Connects to the data pin of the right pointing perimeter sensor.
16	Connects to the data pin of the right angled perimeter sensor.
17	Connects to the data pin of the front pointing perimeter sensor.
18	Connects to the data pin of the left angled perimeter sensor.
19	Connects to the compass data
20	Connects to the compass clock
21,22	Connects to the Right Motor (and one connects to the right servomotor header)
23,24	Connect to power connector though the switch

Power is routed appropriately to all devices.
Capacitor pads are provided throughout the board.
Contact us if you have questions.

Testing the PCB
The RROS has so many capabilities and so many possible configurations, that it is impossible to just give you a program that tests everything. That is, after all, why we provide this lengthy manual.

We do want to get you started though. Let's examine the short program in Figure D-11. The two include files MUST be in the same directory as the program, or you must provide the full path to the file instead of just the file name as we did here. **Note**: If you wish, you can MERGE these files permanently into your programs (from the FILE menu options in RobotBASIC) instead of including them. These files are provided in the SamplePrograms folder in the ZIP file download for the RROS.

Either way, you will have to edit the INITIALIZATION routine that you use (we have provided several examples for you in the InitializationRoutines.bas file). At the very least, you should change the PortNumber in the rCommPort command so that it matches the BlueTooth port YOU are using. Note: This program is in the SamplePrograms download.

```
/******************************************************************
* Both of the include files (below) are simple RobotBASIC programs *
* and may be opened and edited. You MUST modify the parameters to *
* match what YOUR robot needs. For example, change the PORT # of  *
* the virtual serial port used for your Bluetooth link. You should *
* also setup what sensors you are using or problems may occur. If *
* your Init file says you have a compass, and you don't, then the  *
* RROS will hang up trying to communicate with it. You can make   *
* your own custom Init files that handle robots with any motor and *
* sensor configurations. Depending on the weight and batteries     *
* used by your robot, you may also need to modify the various      *
* speed parameters.                             *
*                              *
* Once you create the appropriate Init file, your robot can be    *
* controlled as easily as the RobotBASIC simulator. Read the RROS *
* manual (free download) for more detailed information.         *
*                         *
* The following demonstration program will move the robot forward *
* a little, then back, before going into a loop to continually    *
* print the values for the FEEL, BUMP, and RANGE sensors. Move    *
* your hand around the robot and watch these readings change.     *
******************************************************************/
#include "RROScommands.bas"
#include "InitializationRoutines.bas"
gosub InitRROScommands // found in RROScommands.bas
gosub InitRB9robot // use the initialization routine for YOUR robot
rForward 50
rForward -50
rTurn 90
xyString 100,80,"rFEEL   rBUMPER    rRANGE"
while 1
 rForward 0 // any 'movement' will update sensor readings
 xyString 100,100, rFeel();""; rBumper()&14;"";rRange();""
 // the &14 above is because we have no rear bumper on our RB-9
wend
end
```

Figure D-11: This test program moves the robot forward and backward, then continually displays some of the sensors.

Finally, we know our RROS system is a totally new paradigm, and that it can be confusing for those new to this way of building robots. For that reason, we are working on a new book composed of many exercises that take the reader slowly from building a RROS-based robot to simple programs like the one above to complex examples that prepare you for your own experimentation. Refer also to Appendix F for another example.

Assembling the RB-9 Chassis

When we were building the RROS system, we looked for suitable chassis that were reasonably priced and sized, with wheel encoders. For the most part, we could not find anything we felt was of good quality for less than several hundred dollars. We ended up building numerous custom prototypes to test the RROS routines.

Based on our emails though, most people do not have the time or tools to build their own custom chassis. For that reason we began looking for a manufacturer that could build a chassis that we designed. It took a while to fine-tune the design so that we were happy with it, but we think we have a real winner. Figure E-1 shows a fully assembled RB-9 (RobotBASIC, 9 inch diameter robot).

The picture also the optional RB-PCB with ultrasonic sensors to create the VSS (Virtual Sensor System). The PCB is by far the easiest way to create a RobotBASIC compatible robot. Let's look at some of the features of the RB-9 kit.

It has two 9 inch diameter plates made from laser-cut acrylic with spacers that allow our PCB to mount between the plates. This means the top is available for mounting an arm or other apparatus. If you wish to use digital IR sensors instead of ultrasonic or analog IR you will have to run cables from the board to the sensors and mount them near the edges of the plates (numerous holes have been provided for those wishing to experiment.

The top plate has various hole patterns to accommodate arduino boards and Parallax's BOE plus others you can use for custom boards. New holes can be drilled, but you must drill slowly with a sharp bit and take care to minimize pressure so as to not shatter the acrylic.

Notice the slots on the bottom plate. These allow line sensors or other sensors of your choosing to be easily added. Most of the other desired sensors (perimeter sensors, battery monitoring, beacon detection, compass, etc) are all on board the RB-PCB minimizing cables for many configurations. Even the Bluetooth adapter is on board.

The connections for the motors on the RB-9 (both the power and wheel encoder connections) plug directly into the board – the appropriate sockets are included in our optional parts kit for the PCB.

Figure E-1: The RB-9 chassis has been designed to provide
maximum performance for a reasonable price.

The motors on the RB-9 can be driven by the directly by the RROS chip which means you do not need an external driver. The SR04 sensors work as long as they have the hoods as described in Appendix D. The PCB also has sockets for the more expensive Ping sensors (which do not need hoods because they can be individually triggered).

NOTE: The RB-PCB will also mount on the Rover 5 (tank robot) and on the Magician chassis when using digital sensors (a very cost effective way to build a starter robot).

When you receive the kit, it is packed in bubble wrap and contains numerous small bags of hardware and brackets. As you proceed through the assembly process, we hope you appreciate some of the finer points of the design.

The first thing you need to do is remove the protective paper from all the plates and even from the small plastic circles contained in one of the bags. This is more easily done by using a sharp knife to start and edge, then peeling it away as shown in Figure E-2.

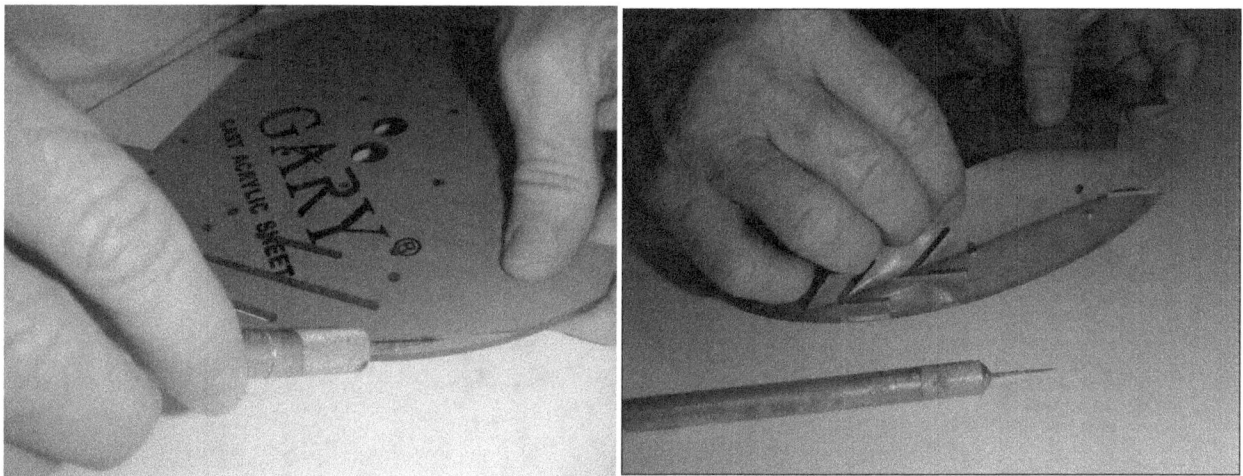

Figure E-2: Remove the protective paper from all plastic parts.

Mount the outside motor brackets and motors as shown in Figure E-3 using the nuts with the plastic centers. Notice the bottom of the bracket points to the outside of the plate. It is important to realize there is a LEFT and a RIGHT bracket to allow the motor to mount securely – when you try to place the motor against the bracket it should be obvious which is which. To secure the motor to the bracket, remove the two screws on the end of the motor and replace them with the larger screws provided (there should be four of these in one of the packets). The packet that contains more than four screws will be used for the spacers.

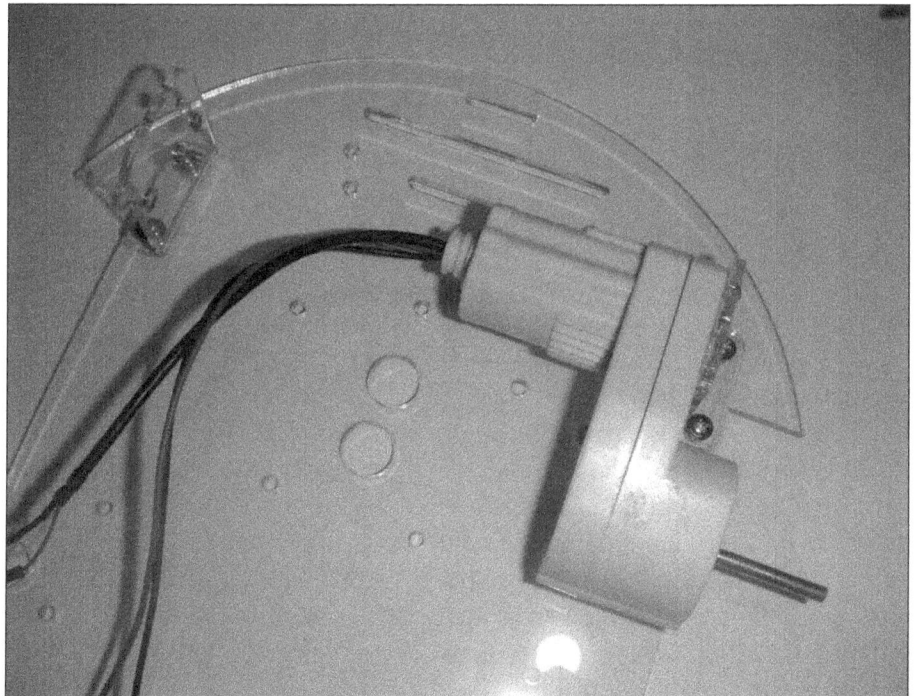

Figure E-2: Mount the outside motor brackets as shown.

Before mounting the second motor place the brackets with the teeth on BOTH motors – there will not be room to do so after the second motor is mounted. Figure E-3 shows both motors mounted properly.

Take care to not kink the wires from the motors. The power wires will feed to the back of the robot and the encoder wires will feed to the front. The encoders have a black (ground) and red (power) that will connect to the ground and power pins on the PCB (polarity is marked on the board). The other two wires are for a quadrature encoder. The RROS only need one of these wires (either will work fine). It should connect to the third pin on the encoder sockets.

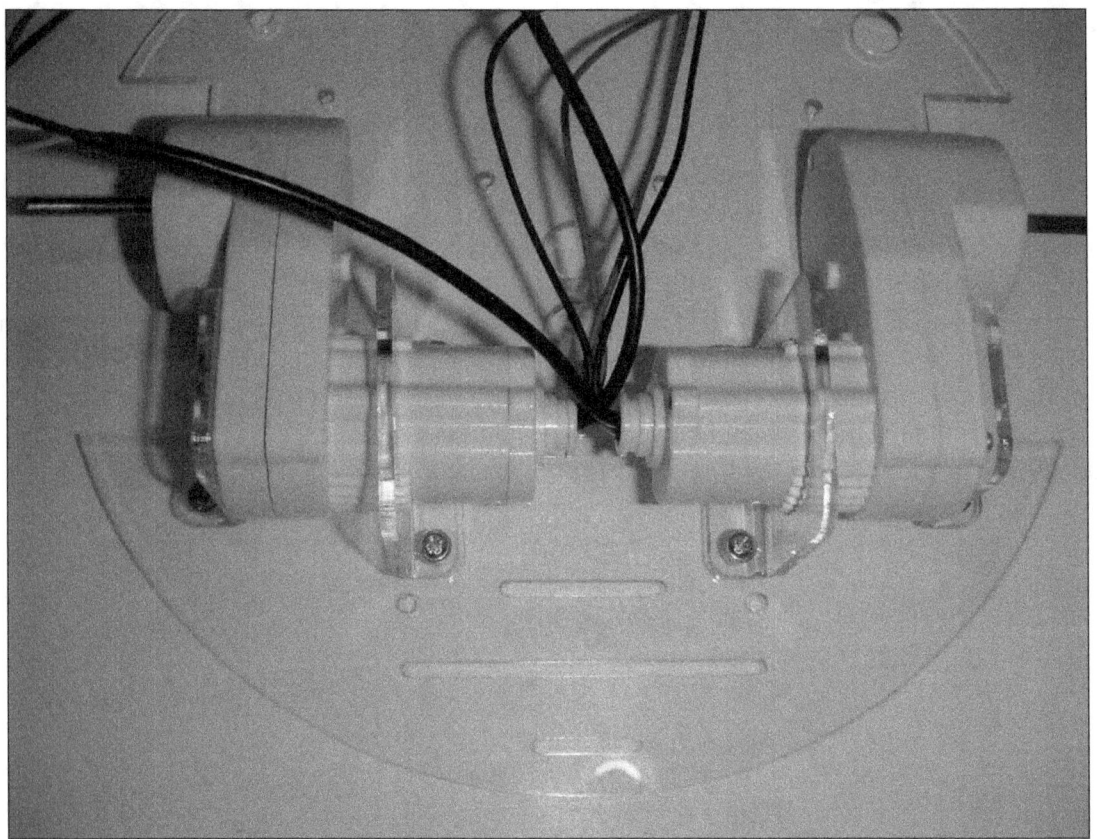

Figure E-3: Both motors mounted.

With the motors installed, we can turn our attention to the wheels. We wanted wheels with excellent traction, but nothing in a reasonable price would mount securely to the shaft. The final design works great but it can be a little confusing so read the following instructions carefully.

There are two round plates for mounting each wheel. One of the plates has three six-sided holes – this plate goes on first. Notice that the center hole in the plate matches the shaft. Slide it on the shaft until it is snug against the motor. IMPORTANT – pull softly on the shaft so that it is as far extended as possible while keeping the round plate against the motor. Secure the plate in this position by using the Allen Wrench (provided) to screw one of the tiny set screws into the hole beside the shaft. This forces a tight fit to the shaft. Make sure you keep the plate parallel with the side of the motor while you install the set screw. Again, make sure the motor shaft is pulled out and the plate is as close to the motor as possible. See Figure E-4.

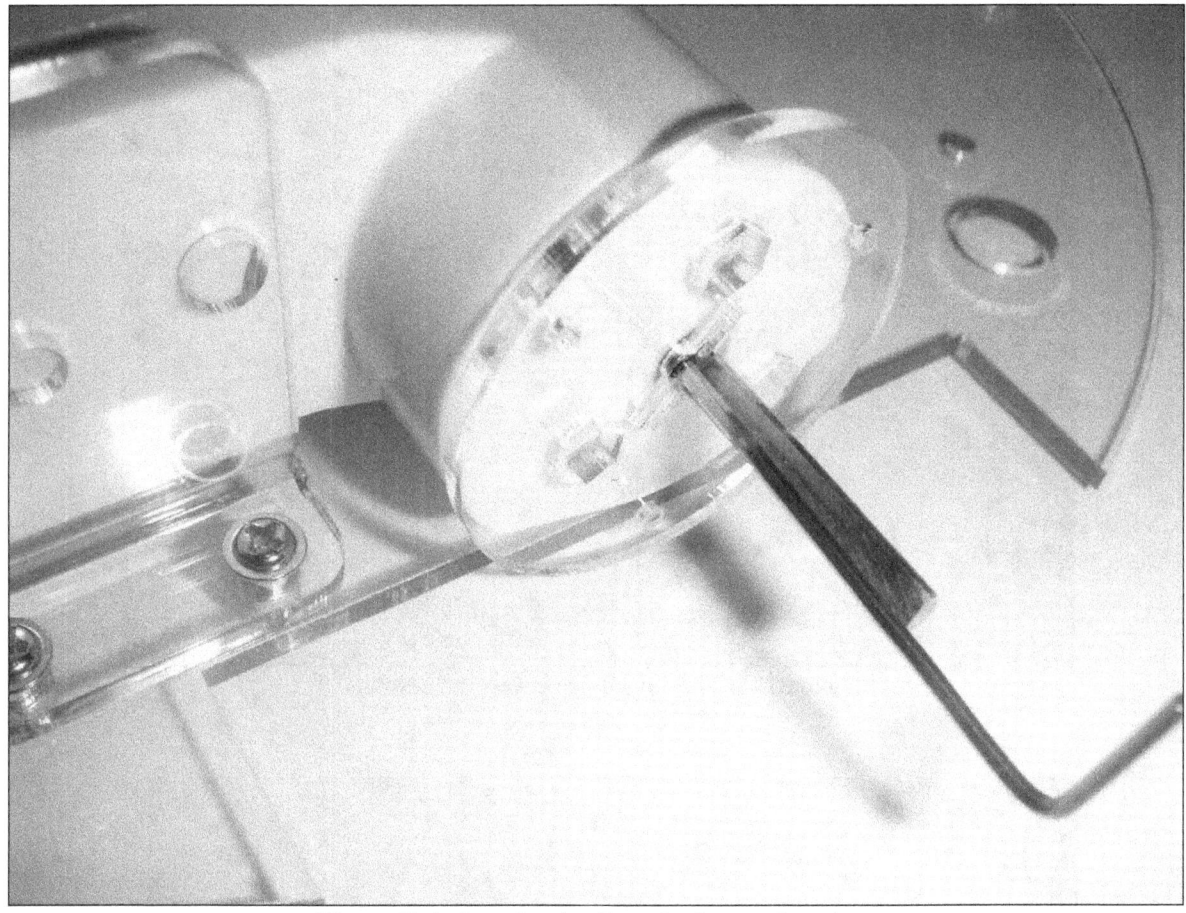

Figure E-4: Securing the first wheel mounting plate.
Notice the Allen Wrench inserted into the set screw.

You only need four set screws total, but the kit probably has an extra one or two. Some small parts are easy to loose, so we have tried to provide extras, just in case.

Next, turn the motor up so the shaft points toward the ceiling and place three nuts into the hex holes (regular nuts, not the ones with plastic centers). Add the second plate (this one has round, not hex holes) to cover the nuts as shown in Figure E-5. Secure it to the first plate with 3 of 6 tiny screws provided (the other three are for the other wheel). Finally, insert another set screw with the Allen Wrench.

There are a number of small plastic donut shaped pieces needed for the next step. Thread 2 donuts on a bolt and screw it into one of the hex nuts as shown in Figure E-6. This creates 3 round "knobs" on the plate that will fit in the holes on the wheels. Add two more knobs.

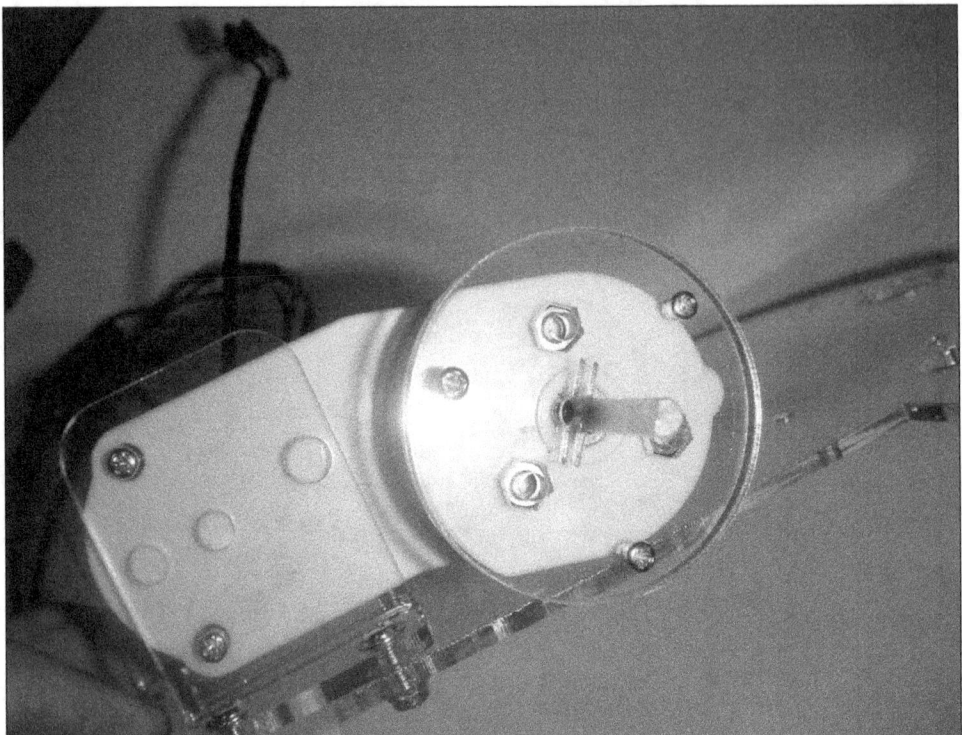

Figure E-5: The second plat covers the hex nuts.

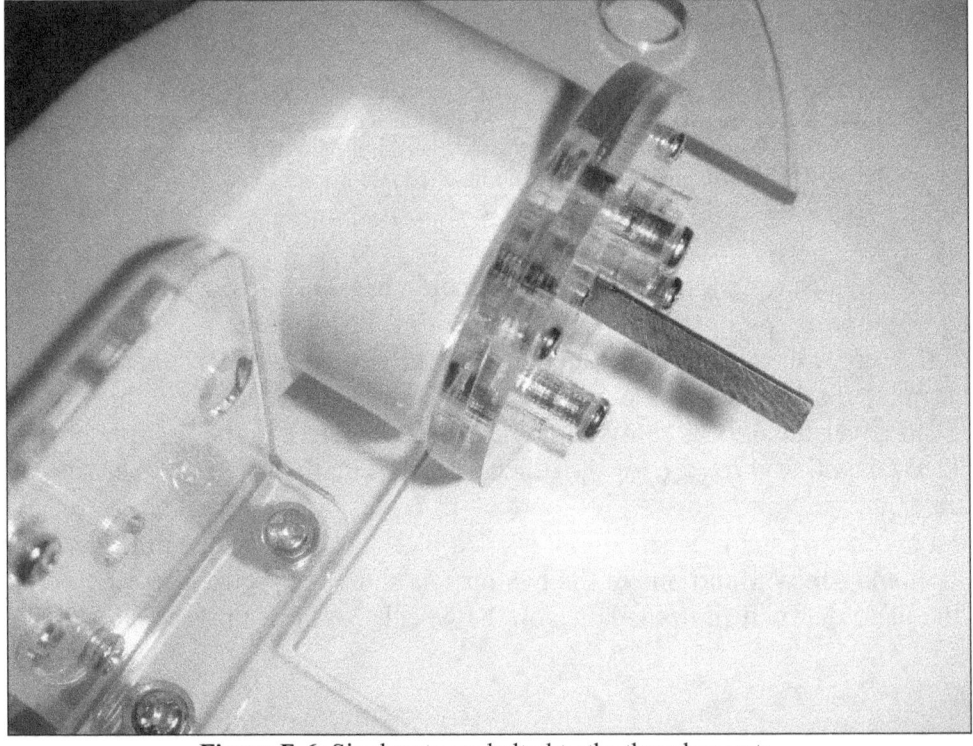

Figure E-6: Six donuts are bolted to the three hex nuts.

Next, mount the wheel against the plate assembly as shown in Figure E-7. The donuts should fit slightly INSIDE the holes in the wheels. Secure the wheel with a collar (use one of the tiny set screws pushing against the flat side, to hold it in place).

Figure E-7: The holes in the wheel should match the donuts mounted on the plates. A collar with a set screw holds the wheel in place.

Perform the same operations on the other wheel and you will have the assembly shown in figure E-8. Thread the power wires under the motors to the rear with the encoder wires forward.

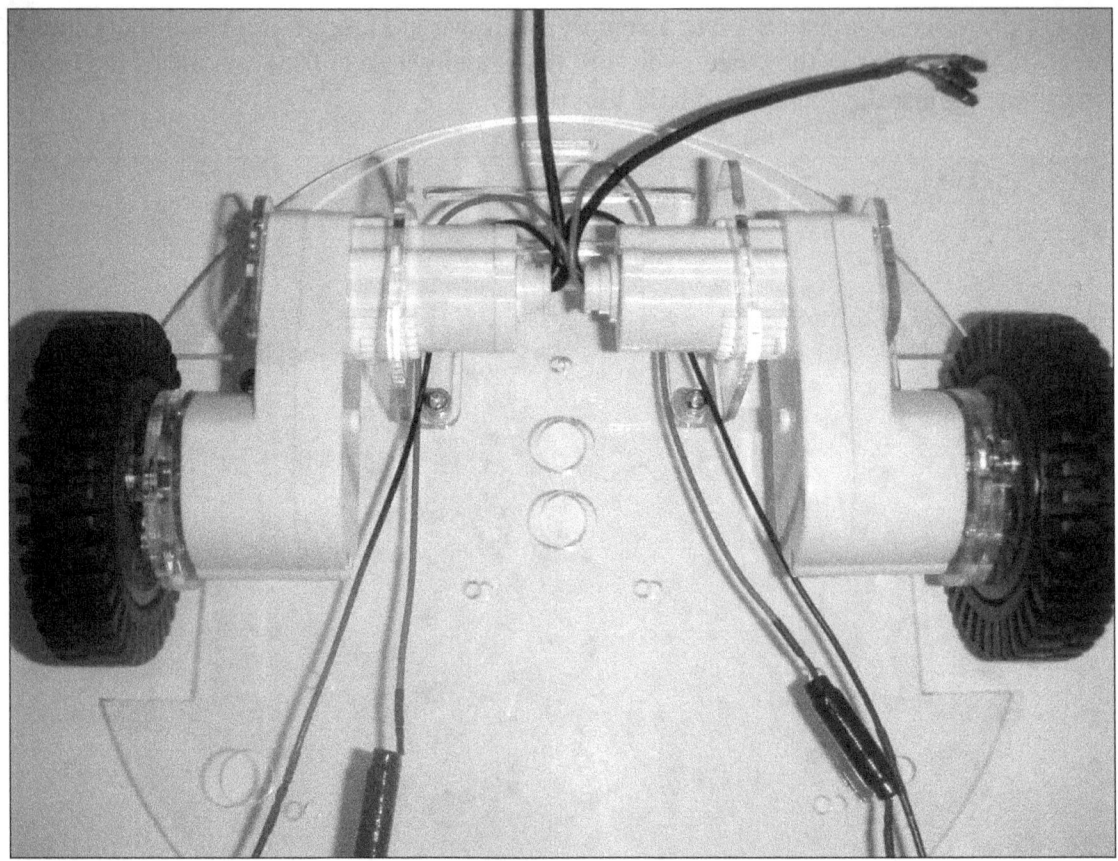

Figure E-8: Both motors mounted to the bottom plate.

Figure E-9: The caster assembly is raised slightly with 2 washers.

The two pieces of the caster assembly must be connected together using two screws that insert from the top and tap into the black ball housing. Do NOT over-tighten and strip the plastic. Mount the caster as shown in Figure E-10, using the nuts with plastic centers. Notice the extra washers to slightly lift the caster (this makes the chassis level with the floor).

Two spacers can be mounted just to the rear of the two large holes in the bottom plate. Small pieces of rubber tubing (provided) can be placed over the spacers making a snug fit for the 6V Gel-cell battery we recommend for the chassis. If you are using another battery, you can eliminate these spacers or move them to a better position.

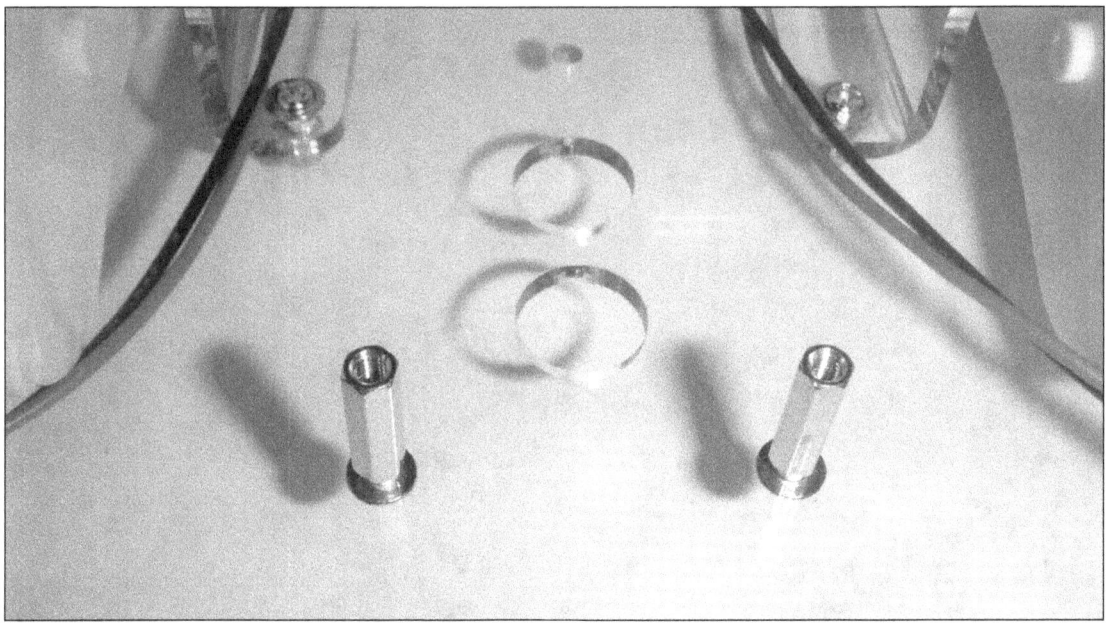

Figure E-10: Spacers serve as backstop for a 6V Gel-cell battery.

The next step is to mount the spacers that support the top plate. **Note:** Depending on the nature of your experimenting, you may choose to leave the top plate off (temporarily) so you can easily connect and disconnect various cables or devices to the PCB.

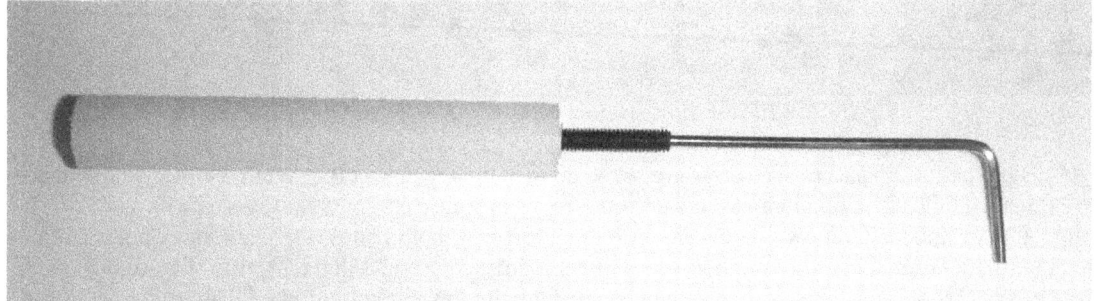

Figure E-11: The small spacers connect together with the longer set screws (use Allen Wrench to insert).

There are 2 long plastic spacers that support the top plate in the rear. You may use any holes that are appropriate for your needs, but you will find holes that are just wider than the recommended

Gel-cell. Secure the two longer spacers to the bottom plate using the screws provide… careful not to over-tighten.

There are three sets of short spacers on the front that support the PCB between the two plates. Each set of spacers is connected together with long set screws as shown in Figure E-11. It is suggested that you create threaded holes in one end of both the top and bottom section by screwing in the set screw as shown in the Figure. Back it out of the first one, then install it into the other. This will make it easier when you screw the two spacers together (using your fingers).

Once you have the short spacers prepared, secure the three bottom spacers (without the set screws) to the bottom plate. Use the holes right in front of the motor brackets and the hole where the two motors come together in the center.

Figure E-12: The short spacers support the PCB and the top plate.

Place the PCB on the lower spacers and screw the set screws (in the top spacers) through the PCB and into the previously threaded holes in the bottom spacers until the board is secure as shown in Figure E-12.

Add the top plate and, assuming you are using holes that match in both plates (there are many holes that do not – holes used to mount Arduino boards etc.) then you will get a finished product that resembles Figure E-1.

There are holes for various sizes of toggle switches on both the top and bottom plate making it easy to add a second ON/OFF switch since the on-board slide switch is not easily accessible when the top plate is secure. Just leave the slide switch in the ON position if you use a second switch.

The motor power lines connect to the motor connections on the rear of the PCB or you can connect them to your own circuits if you are not using ours. The encoder lines should feed from the front and connect to the sockets labeled L.CNT and R.CNT (Left and Right counters).

Of course, many hobbyists will customize the chassis, but this Appendix should serve as a guide for them.

See the end of Appendix D for a short test program to test the RB-9 Chassis with the RB-PCB. Also, See Appendix F for another example.

NEW BOOK

We are considering a new book that provides a series of hardware/software experiments that helps the reader learn about programming and the simulator and the RB-9 robot or perhaps Arlo. We would love to work with a local school and create experiments that are appropriate for every grade level from 5-12. While it would be aimed at schools, such a book would be great for hobbyists at all levels too. Any school system interested in participating should contact us.

Example with Practical Considerations

We know that the RobotBASIC way of building a robot is a new paradigm for many hobbyists and we understand that many aspects of the RROS can be confusing at first. For that reason, we are including this simple example to help demonstrate some of the practical aspects of moving a simulation-based program to a real-world robot.

We need an example application that can demonstrate several principles, yet one that can be easily understood and followed even by beginners. We chose to have the robot rotate at its position and look for nearby objects. When a relatively large object is found, the robot should move to the object.

We will start by showing a simple means of achieving this goal using the simulator. Refer to Figure F-1 as you read the following discussion. The program will start by using an **Init** routine to draw some objects and to position the robot at the center of the screen. Objects will be detected using only a front-pointing range sensor, which can be read using the function **rRange**().

The main **repeat** loop causes the robot to continue searching until a suitable object has been found.

```
MainProgram:
 gosub Init
 repeat
  done = false
  repeat  // look for a nearby object
   rTurn 1
  until rRange()<150
  delay 2000  // delay when object found
  A1=rCompass()
  repeat
   rTurn 1
  until rRange()>150  // keep turning until past object
  delay 2000  // delay when object no longer seen
  A2=rCompass()
  TotAngle = A2-A1
  if TotAngle>180 then TotAngle = 360-A2+A1
  if TotAngle >45  // large object found
   // face object
   rTurn -TotAngle/2
   // go to object
   while rRange()>20
    rForward 1
    done = true
   wend
  endif
 until done
end
```

```
Init:
   gosub DrawObjects
   rlocate 400,300
   rSpeed 10
return

DrawObjects:
   Linewidth 10
   SetColor Blue
   Line 300,340,250,250
   circlewh 400,400,10,10
   circlewh 500,300,30,30
   Line 600,500,700,450
return
```

Figure F-1: This program allows the simulated robot to find a large object that is nearby.

Inside the main loop, a second **repeat** loop turns the robot to the right until it detects an object closer than 150 pixels. When one is detected, the robot pauses for two seconds so you can see that it has detected an object. At this time, the current compass heading is stored in the variable **A1**.

Next, another loop continues turning the robot until the object is no longer seen. Again the robot pauses, and records the current heading in **A2**. The difference in these two headings gives an indication of the size of the object detected. For example, if the first reading was 50° and the second was 85°, then the view of the object spans 85-50 or 35° degrees. We will have the robot look for objects that span more than 45°. **Note**: Using only the 45° parameter to identify large objects is a simplification made to keep the program easy to follow. In a more robust program, many parameters should be considered. For example, the program should take into account how far away the object is, because the same sized object will have a smaller angular span when it is further away.

It is also important to realize that the formula (**A2-A1**) for the span does not work if the object is detected in a northern position. An example can illustrate this point. Suppose the object is detected at 320° and spans until 35°. This span is actually 75°, but the formula indicates 285° making even small objects appear to be greater than the 45° limit. This problem can be easily corrected by checking to see if the original value for the total angle is greater than 180°, meaning the span must be crossing the due North reading. When this happens, the total angle is calculated as **360-A1+A2**, thus giving the proper value.

If the total angle is greater than 45° the program assumes an appropriate object has been found and commands the robot to turn back to the left an amount equal to ½ of the span – making the robot face toward the middle of the object. The robot then moves forward until it is close (20 pixels) to the object.

This is a very simple behavior, yet it allows the robot to carry out a very specific mission. It should be pointed out that if no large objects are nearby, that the robot will continue to rotate. Implementing a solution to this problem is left for the reader. It could be as easy as monitoring the system clock and only allowing the robot to perform its actions for a limited amount of time.

Moving to the Real World

Thinking about the practical considerations of moving this program from the simulator to a real RROS-based robot is what this Appendix is about.

One of the first things to consider is how to initialize the real robot instead of using a simple **rLocate** to create the simulated one. We will use the value of a variable called **REAL** to make this determination. When **REAL** is **true**, we will initialize the real robot. If it is **false**, then the simulation will be used. More on this shortly.

Another thing to consider is the distance measured by **rRange**(). The simulator reports this distance in pixels while the RROS generally reports the distance in increments of ½ inch. You could easily have the program check for different distances based on the value of **REAL**, but there is another alternative.

If you use the **rCommand**, **SetRobotDiameter** to set a non zero value for your robot's diameter (in ½ inch increments), then the RROS will report **rRange**() readings in pixels just like the simulator. If you are using the RB-9 chassis (which is 9 inches in diameter) for example, you would inform the RROS of its diameter with this command.

  rCommand(SetRobotDiameter, 18)

Since the RROS now knows how large your robot is, it can convert the **rRange**() parameters to equivalent pixel readings. Remember, the simulated robot is 40 pixels in diameter, so if an object is 9 inches from the robot, it can also be considered to be 40 pixels away. These readings will never be as accurate as those of the simulator, but they do make it easy to write programs so that real robots respond in a similar manner to the simulator. **Note**: Earlier versions of the RROS did not have the **SetRobotDiameter** command so if your version does not work, or does not work properly under all conditions, you can return the chip and have it reprogrammed with the latest version for free.

Having **rRange**() report its readings in pixels is not always enough to ensure compatibility though. The simulated ranging sensor detects objects along a straight line extending outward from the robot. This is very similar to the functionality of an IR ranging sensor and if you use one on your real robot, the readings will be very similar to that of the simulation.

If your robot uses an ultrasonic ranger though, the detection area will be a cone rather than a straight line. This means that the real robot, as it turns, will detect objects well before simulated one. This will make objects appear much larger than they actually are (when compared to the readings of the simulator). This is easy to fix by adding a value to the desire readings for the span angle. The amount to add depends on the size of the detection cone associated with your sensors, but a value of 30° or so is not unreasonable.

It is also important that the real robot move at a relatively slow speed. Otherwise, the robot might move well past the angular position where a reading was taken by the time it decides to stop. It is also vital that you calibrate the compass periodically in order to get accurate readings.

All of the above considerations are easy to account for. They are mentioned here because many first-time RROS users assume that the real robot will respond exactly like the simulation. With proper calibration of the real robot and appropriate consideration of the differences between your robot and the simulated robot, an acceptable correlation is possible. If your real robot is not responding as you expect, examine all of the relevant parameters to find criteria that might be causing problems.

There are many differences between a real robot and the simulation. Many of these differences (as well as possible solutions) are discussed throughout this text so read it carefully if you are having problems.

Figure F-2 shows a modified version of Figure F-1 that will work with both the simulation or a real robot based on the value of the variable **REAL**. This program is included in the Sample Programs download for the RROS.

In the original program, the simulated robot always starts with a due North orientation. This program locates the simulation in a random orientation to make it more like a real robot. In addition, both robots are turned to North before the program continues. The **TurnToHalfAngle** command requires the init file to setup the TurnTo max and min speeds to work properly. These speeds should be set experimentally for your robot to allow it to create accurate TurnTo movements. Note: These speeds will generally be fairly large because the RROS must read the compass while the robot is moving – thus creating a much lower effective duty cycle for the DC motor control.

```
#include "RROScommands.bas"
gosub InitRROScommands
/*****************************************************************
* Normally, you would #include the InitializationRoutines.bas file here *
* and then call the appropriate routine for your robot. The standard     *
* RB-9 init routine does not include the compass (because it would cause *
* the program to 'hang' if no compass is present. Rather than change     *
* that standard routine, this program includes a copy of the RB-9 init   *
* routine with the compass included. It is suggested that you modify      *
* the initialization routines you use so they match your specific         *
* hardware.                                                               *
*                                    *
* This program rotates the robot until it finds a large object. When one *
* is found, the robot moves to it.                                       *
*****************************************************************/

MainProgram:
 REAL=True // change to true if a real robot is being used
 gosub Init
 //rCommand(CalibrateCompass,0)
 gosub FaceNorth
 repeat
  done = false
  repeat  // look for a nearby object
   rTurn 1
  until rRange()<150
  delay 2000  // delay when object found
  A1=rCompass()
  repeat
   rTurn 1
  until rRange()>150  // keep turning until past object
  delay 2000  // delay when object no longer seen
  A2=rCompass()
  TotAngle = A2-A1
  if TotAngle>180 then TotAngle = 360-A2+A1
  if TotAngle >45  // large object found
   // face object
   rTurn -TotAngle/2
   // go to object
   while rRange()>20
    rForward 1
    done = true
   wend
  endif
 until done
end

Init:
 if REAL
  gosub InitRB9robot // use a routine for your robot (or modify this one)
 else
  gosub DrawObjects
  // init the simulator
  rlocate 400,300,random(360)
  rSpeed 10
 endif
return

FaceNorth:
 if REAL
  rCommand(TurnToHalfAngle,0)
 else
  while rCompass()!=0
   if rCompass()>180
    rTurn 1
   else
    rTurn -1
   endif
  wend
 endif
 delay 3000
return

DrawObjects:
 Linewidth 10
 SetColor Blue
```

```
 Line 300,340,250,250
 circlewh 400,400,10,10
 circlewh 500,300,30,30
 Line 600,500,700,450
return


InitRB9robot:
 // modify everything as necessary to match your hardware
 rCommport 47 // change to the virtual port used for your Bluetooth device
 SetTimeOut 50000
 rlocate 0,0 // 255,0 if communication delay is needed
 rCommand(MotorSetup,SMALLDC+ENCODERS)
 rCommand(SensorSetup,SRO4+FIVERANGE+HMC6352)
 rCommand(SetMotorRamp,5)
 rCommand(SetClicksPerDiam,160)
 rCommand(SetClicksPer90,110)
 rCommand(SetReducForwRight,0)
 rCommand(SetReducForwLeft,15)
 rCommand(SetReducBackRight,0)
 rCommand(SetReducBackLeft,0)
 //rCommand(SetRotationTime,18)
 //rCommand(SetMoveTime,12)
 rCommand(SetSpeed,35)
 rCommand(SetSlowDownSpeed,30)
 rCommand(SetSlowDown2,25)
 rCommand(SetCCdivisor,3)
 rCommand(SetBumpDist,10)
 rCommand(SetProxDist,15)
 rCommand(EnableCounters,1)
 rCommand(SetRobotAngle,90/2)
 rCommand(SetTurnStyle,100)
 rCommand(SetRROStimeout,15)
 rCommand(SetTurnToMin,45)  // turn to speed must be FASTER because of
 rCommand(SetTurnToMax,70)  // the delay to read the compass
 rCommand(SetRobotDiameter,18) // calibrates rRange to pixel measurements
return
```

Figure F-2: This program is equivalent to Figure F-1 except that it will control either a real robot or the